THE LANDSCAPE OF INDUSTRY

The Landscape of Industry is a comprehensive study of one of the most significant sites in the development of early industrial culture. The Ironbridge Gorge, still marked by the Iron Bridge, is now recognised by UNESCO as a World Heritage Site.

Judith Alfrey and Catherine Clark bring buildings and landscapes together in an integrated survey which sets the extraordinary events at Ironbridge in their chronological and spatial context. In short, they explore the process of industrialisation to show how it may be *read* in the environment. The book considers a range of material evidence, combining archaeological appraisal of the industrial landscape with analysis of its settlement patterns and buildings. Its inter-disciplinary approach establishes a method for the study and analysis of complex industrial landscapes which may be applied to many other areas.

This book will be essential reading for all who are concerned with the investigation and preservation of industrial sites throughout the world.

Judith Alfrey is a building historian who now works as a historic buildings consultant. She is co-author (with Tim Putnam) of *The Industrial Heritage* (Routledge 1992).

Catherine Clark is Inspector of Ancient Monuments with English Heritage and formerly Conservation Officer with the Council for British Archaeology.

THE LANDSCAPE OF INDUSTRY

Patterns of change in the Ironbridge Gorge

Judith Alfrey and Catherine Clark

ROUTLEDGE
London and New York

First published 1993
by Routledge
11 New Fetter Lane, London EC4P 4EE

Simultaneously published in the USA and Canada
by Routledge
29 West 35th Street, New York, NY 10001

Typeset in 10 on 12 point Garamond by
Selwood Systems, Midsomer Norton

Printed and bound in Great Britain by
Biddles Ltd, Guildford and Kings Lynn

British Library Cataloguing in Publication Data
A catalogue record for this book is available from the British Library

Library of Congress Cataloging in Publication Data
Alfrey, Judith
The landscape of industry: patterns of change in the Ironbridge Gorge / Judith Alfrey & Catherine Clark.
p. cm.
Includes bibliographical references and index.
1. Industrial archaeology – England – Ironbridge (Shropshire)
2. Mines and mineral resources – England – Ironbridge (Shropshire) – History. I. Clark, Catherine (Kate) II. Title.
T26.G7176 1993
609.424'56–dc20 93–20280

ISBN 0-415-03319-5

CONTENTS

LIST OF ILLUSTRATIONS

PREFACE

The literature of the landscape is a vast and continuously expanding one. But perhaps because of literal definition of landscape as 'scenic', there has been relatively little written on the landscapes of industry. The industrial landscape may seem to comprise waste heaps, ugly buildings, squalid houses, and is often dealt with in a perfunctory manner, as being too complex, or too destructive. Yet in shaping what we see around us, industrialisation – not just the Industrial Revolution but also the years preceding and succeeding it – has been of paramount importance.

This book has set out to study such a landscape in one small area, the Ironbridge Gorge. Its research and preparation was funded by the Nuffield Foundation, and based at the Ironbridge Institute. The original brief for the survey was to develop new ways of undertaking field survey in industrial areas, as well as to provide a data base for the Gorge itself.

The survey was special for a number of reasons – because it covered such a small area, because it dealt with an area with a strong historical tradition and rich documentary sources, and because we were able to combine the study of standing buildings with other forms of landscape evidence. What we hope we have been able to do as a result is to begin to use landscape evidence critically, as a historical source in its own right.

If the landscape is a historical source, it is also a valuable resource. As ecological environment, the landscape is something we now cherish and care for. But there is also a human element to that environment – an inherited historical landscape which surrounds us, and shapes our activities. Those decisions which might affect the ecological environment will also affect this historical landscape. We hope that by exploring the potential of the landscape for the study of history, as well as considering its role in shaping history, we have set out a case for a more considered and integrated approach to its conservation.

Judith Alfrey
Catherine Clark

ACKNOWLEDGEMENTS

The research and preparation of this book, based at the Ironbridge Institute and carried out between 1985 and 1988, were made possible by a grant from the Nuffield Foundation; we would like to express our thanks to the Foundation for this funding, and to Anthony Tomei for his particular help and encouragement. We have also been helped throughout by the staff of the Ironbridge Institute and Ironbridge Gorge Museum Trust, and would especially like to thank Michael Stratton and Barrie Trinder, John Powell and Michael Vanns, Janet Markland and Carol Sampson. Many other people have helped us in many ways, during research and writing, but we are particularly grateful to all those people in the Ironbridge Gorge without whose co-operation the fieldwork for this study would not have been possible.

Judith Alfrey
Catherine Clark

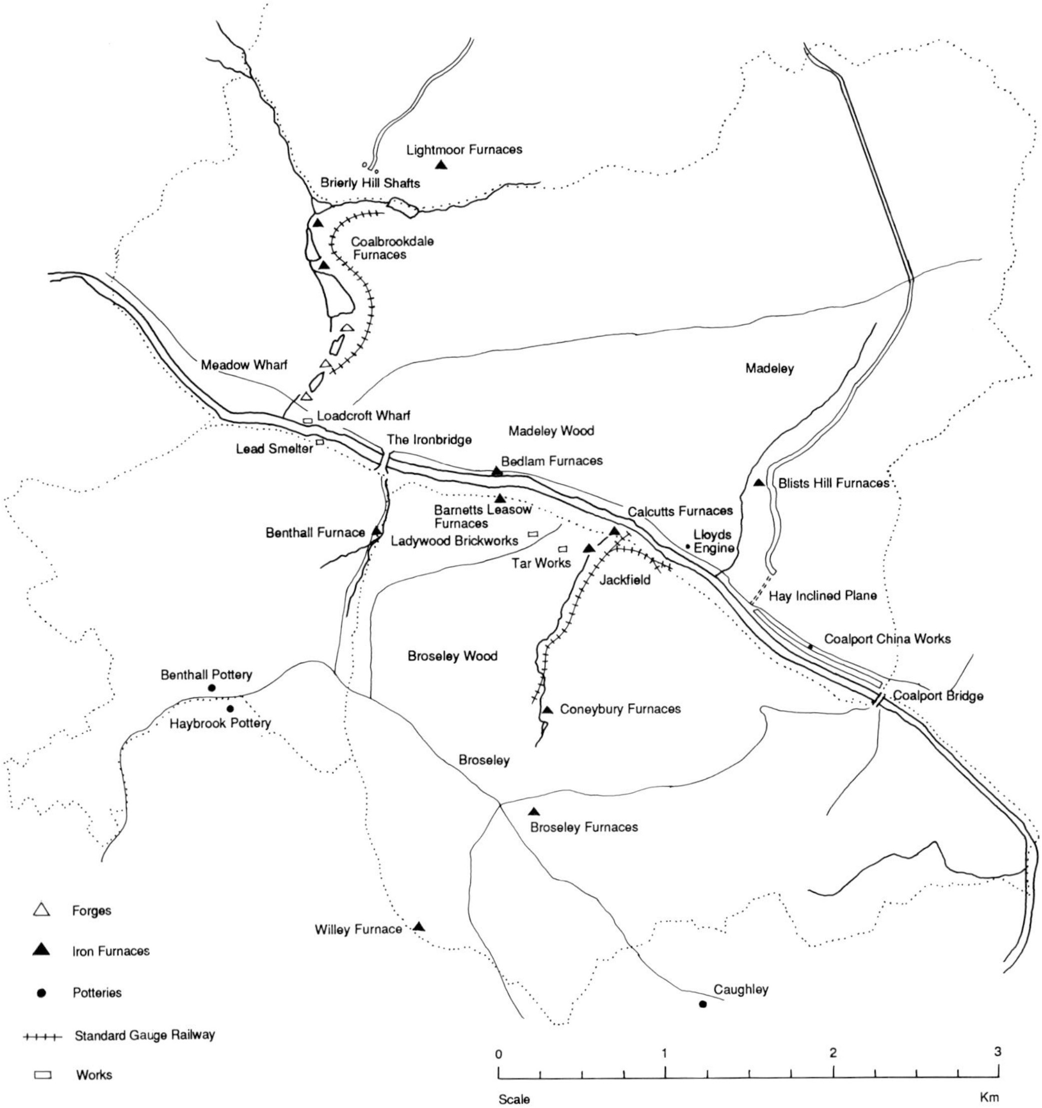

0.1 The Ironbridge Gorge: eighteenth-century sites mentioned in the text (Shelley White)

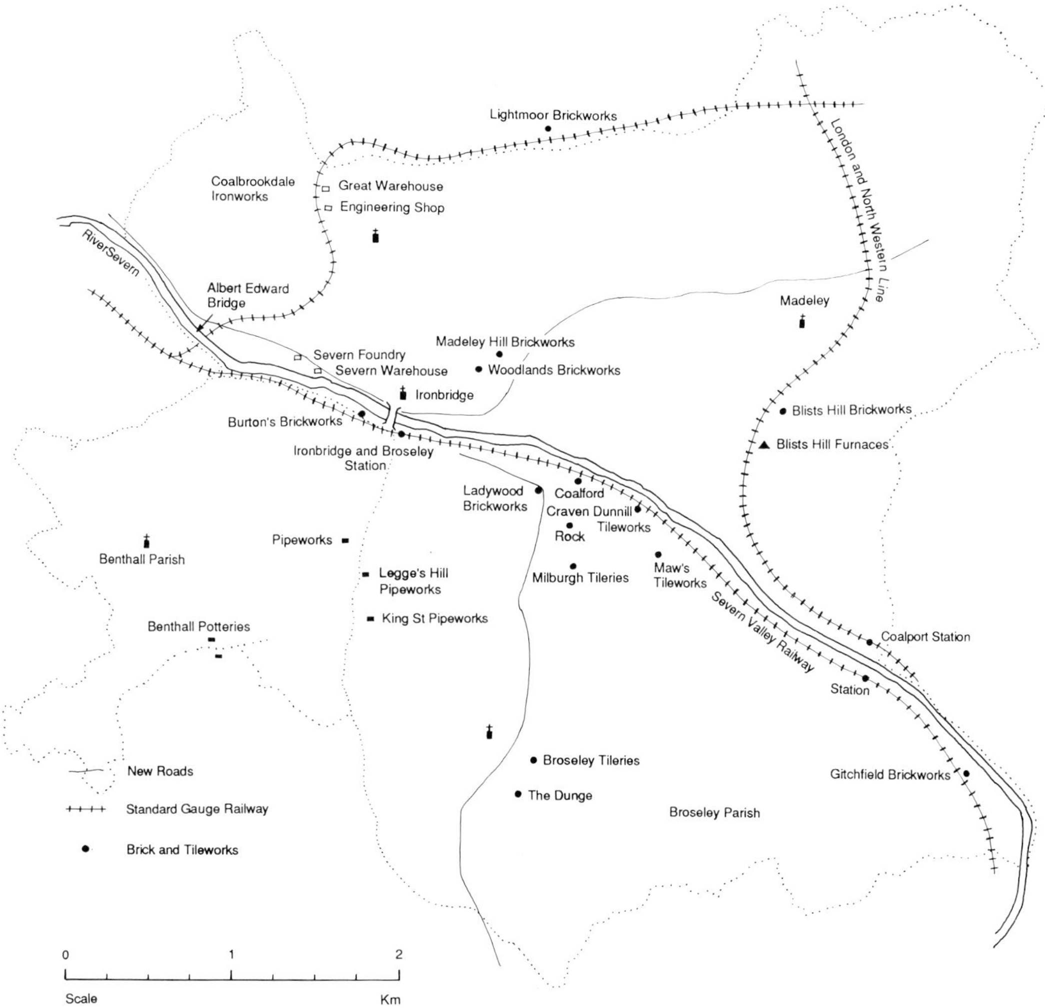

0.2 The Ironbridge Gorge: nineteenth-century sites mentioned in the text (Shelley White)

1

THE LANDSCAPE AS A SOURCE OF HISTORICAL EVIDENCE

The Ironbridge Gorge is often called the 'Cradle of the Industrial Revolution'. It enters history books as the place where Abraham Darby first smelted iron in 1709 using coke and not charcoal as a fuel, thus paving the way for the mass-production of iron. Other technical innovations followed, including the first use of iron rails, and new contributions to the technology of steam engines and locomotives. And it was here in 1779 that the newly discovered properties of iron were demonstrated with a spectacular iron bridge that spans the Severn Gorge, the first iron bridge in the world (fig. 1.1). The Iron Bridge has become a symbol of both the Industrial Revolution and Britain's eighteenth-century industrial achievements.

1.1 'A View of the Cast-Iron Bridge', E. Edgecomb 1786 (SSMT40)

The history of the Gorge illustrates the processes of social and economic change which had their roots in the proto-industries of the seventeenth century, and which continued into the urban economy of the late nineteenth century. The traces of this history are imprinted on the landscape, and this book uses detailed local study to show how the landscape of the Gorge is not just a backdrop for technological achievement, but is also a rich source of evidence for the context and course of industrial change (fig. 1.2).

The book takes as its focus the landscape and communities of the Ironbridge Gorge, and studies their origins, growth and decline. By working within these very narrow parameters, it has been possible to exploit the potential of landscape as a historical source, and integrate physical evidence with other material in an understanding of industrial change, examining not only how industrialisation came about, but also how it was sustained and developed.

1.2 Aerial view of the Ironbridge Gorge (IGMT 1991.2294)

A second and more practical thread to the work was the need for more information for planning decisions. The Ironbridge Gorge is one of less than a dozen World Heritage Sites which relate to the history of industry; it is also a Conservation Area which contains seven Scheduled Ancient Monuments and over two hundred Listed buildings. Yet despite this protection, planning decisions are often made without taking full account of the historical landscape. Elements within the landscape, such as the water-power systems, or early industrial housing, have been lost through failure to recognise their importance, and have not always been acknowledged and reflected in the system of protection. The mechanism for protecting archaeological sites is the Sites and Monuments Record held by the County Council, but it includes only a small proportion of the potential number of sites in the Gorge. Augmenting the Sites and Monuments Record was a prime justification for the work. However, it became clear that decisions needed more than just new archaeological points on the map. It was necessary to identify and present the links between these sites, to understand the historical structure of the landscape. In order to do this, a more complex type of survey was needed.

Landscape evidence

Economic histories of the Industrial Revolution commonly rely almost exclusively on documentary evidence; when physical evidence is used, it is often only to confirm or deny documentary sources, and is not usually used as a primary source in its own right. It is the landscape as a primary source of information which provides the central theme of this book, and its data are all those physical elements which remain or can be recorded, or which can be reconstructed from maps and documents. The principal sources for this study were therefore themselves landscape features – the shape and structure of the land, and the modifications made to it during its exploitation, the physical evidence of industry, buildings, architecture and settlement.

Our use of the term 'landscape' includes the widest range of features – from river to culvert, field to waste-heap, cottage to council-house – but the integrity and coherence of this industrial landscape have been respected by placing each distinct class of evidence firmly in its physical context and drawing out the spatial and chronological relationships between them.

The use of landscape evidence is not new. The evidence constituted by the landscape is familiar territory to local historians, following the pioneering work of W. G. Hoskins,[1] and several other disciplines have as their subject components of the landscape: historical geographers draw inferences from the reconstruction of landscapes, and the spatial dimensions of social and economic conditions; urban historians study urban forms to analyse the development of towns, with a strong emphasis on social patterns; architectural historians take as their data building form; traditional archaeologists have used temporal as well as spatial data to look at evidence for change over time.

Each of these disciplines uses physical evidence – buildings, spatial organisation, archaeological deposits and urban form – but the industrial landscape in its own right can also constitute a class of historical evidence. Although there have been several studies devoted to the identification of the characteristic landscapes and features associated with particular kinds of industrial development, such studies usually select different landscapes to illustrate different stages of development, and do not consider how a single landscape changes. This study aims to develop an integrated approach to landscape study, and does so by bringing together different classes of evidence in the analysis of a single area as it changes through time. It begins to make links between different aspects of the landscape, and between different disciplines and data which can be used in its study.

The data for the study were gathered through an intensive archaeological survey of the Ironbridge Gorge. Unlike many such surveys, we did not confine ourselves to a single period, nor did we stop short at any particular date, analysing with equal attention the modern and ancient landscape. The survey involved the collection of data relating to buildings, plots and the landscape as a whole. This book is a direct product of that archaeological survey, and sets out to explore how landscape study could contribute to a better historical understanding of the Ironbridge Gorge.

Space, time and typology in landscape survey

Why should archaeology, a discipline traditionally associated with earlier segments of the past, be a suitable way of integrating different types of recent landscape evidence? Archaeologists have always had to deal with physical evidence for the past, and often in the absence of written records. The subject has evolved specialised techniques for deriving historical information from artefacts and sites. Three of the fundamental principles which underpin the treatment of such data are by no means unique to archaeology, and taken together they do provide a suitable framework for dealing with landscape evidence.

Recognition of patterns of change through time is basic to archaeology, and is enshrined in the concept of stratigraphy. Stratigraphy is in origin a geological principle, which was applied by excavators to the build-up of layers on a site. The same principle can be applied to landscapes and buildings. The landscape is a palimpsest of past activities, built one on top of the other. By stripping down the layers of this palimpsest, it should be possible to understand what can be seen today. In the same way, buildings change, and it is often only through adaptation that historic buildings remain standing. These changes are not just superficial alterations, but are part of the process of building itself. Recognising this cycle of change forms a key to understanding the historical processes which have shaped the landscape.

The second principle is that of setting information in its contemporary, spatial context. Spatial patterning is a geographical concept, which again is applied in archaeology. The concept of 'context' is the basis of archaeological excavation, in which the information value of a find lies more in where it is found than in what it is.[2] A coin may have no value to an archaeologist unless it can provide information about the context in which it was found. Context in the industrial landscape may be the land around a building, the relationship of a furnace to its water-power system, or the line of a coal outcrop. Identifying 'context' is thus a way of linking the buildings and individual industries with their wider landscape setting.

The third concept is the idea of the typology. The earliest notions about prehistory were based on the recognition that artefacts could be grouped by materials, and this concept has remained a valuable approach to the classification of objects.[3] In the case of a landscape, the objects might be buildings, for example. Buildings themselves are only rarely documented, and the date, builder and occupier are not often known.

But through systematic quantitative and comparative analysis (discussed in more detail below) it is possible to identify significant patterns of variation through time. These patterns can help not only in the dating of local buildings, but also in understanding the social, economic and cultural contexts of their use.

By applying the basic archaeological principles of time, space and typology, we hoped to be able to use landscape data in a more rigorous fashion than had hitherto been possible. But in order to use these concepts to make a coherent analysis of the patterns which we observed in the landscape it was necessary first to refine the field techniques of archaeology to make them more applicable to the particular problems posed by the industrial landscape.

The first difficulty was the traditional archaeological concept of 'the site', which is the usual way of collecting information. There were several problems with the idea of the site understood as a closely delineated point on a map; the first was that it effectively divorced a building or industry from its wider spatial setting. Second, selection of specific sites risked discriminating between categories of information and selecting only the most interesting. For industrial sites where operations are linked to complex networks of transport or power, it can be difficult to draw boundaries. Finally, the idea of a site as a single entity was not easy to sustain where every building or industry seemed to have gone through a complex cycle of reuse and change.

Therefore another means of collecting information was adopted – plot survey.[4] Plots are the basic units in which all land is divided, described and conveyed; they may be of any shape and size and range from a tiny garden to a large area of open heath. They are not just arbitrary divisions of land, but have a historical significance. They reflect continuity in the organisation of land, and the way land is divided; plots can indicate the circumstances in which a building was built, or show changing patterns of agriculture. Plot boundaries may also be important in the recognition of past processes of land development. Plot survey was also a way of ensuring total, rather than selective, coverage of the landscape.[5]

The plots shown on the 1902 Ordnance Survey Map (fig. 4.2) were taken as the basis for the survey. The choice was practical since full coverage was easily available, and also made sense in the field in that the map included the standard-gauge railways which had radically transformed the landscape in the late nineteenth century.[6]

A single plot might contain many different features; during survey numbers were therefore assigned to all identifiable features within each plot, in order to ensure that equal weight was given to different types of information. The archaeological technique of a stratigraphic matrix was used to organise this information chronologically, to sequence and phase complex sites.[7] The value of this technique in a landscape context was demonstrated particularly well in the limestone industry of Benthall Edge, where over a hundred quarries, approach roads and spoil heaps were numbered and recorded. A diagram was constructed showing their stratigraphic relationships, and from this it was possible to identify the different phases of quarrying from the medieval period through to the twentieth century, as well as transport routes, and the type of stone quarried, despite the minimal quantity of documentary evidence. The matrix was thus a way of analysing change through time in the most rigorous manner possible.

Housing as a historical source

The survey which forms the basis for this study also sought to develop the use of housing as a historical source. Settlement patterns and building design can contribute to understanding the process of industrialisation, since they are a direct product of patterns of economic and social change. The location of housing, and the patterns of its development, map out a wider land use, influenced both by the presence of industry and by the organisation of landholding. The distribution patterns, types of development and certain architectural characteristics found in the Gorge themselves provide evidence for the process and structure of industrialisation.

This study used the phasing, distribution and design of buildings as evidence for the history of industrialisation and the creation of an industrial way of life. It aimed to identify those features characteristic of the landscape of industry – its patterns of settlement, its housing types and their significance – and to show how industrialisation took place in the spheres of architecture and building design.

In carrying out this programme of research, the plot-based approach discussed above in the context of industrial archaeology was also used. This meant that building was considered first as a particular land use, and spatial context was given due attention. And since it was the history of the plot, rather than a conventional architectural history, which provided the starting-point for research, no chronological cut-off date was adopted, with the result that any building of whatever age or status was considered. In addition, buildings which no longer survived were recorded wherever possible, using a range of documentary sources. Previous buildings on a site, and the several phases of development characteristic of many of the buildings in the area were also taken into account. The histories of buildings themselves were registered, thus considering not only the moment of construction, but also the patterns of use which modified and conditioned the survival of the buildings in the area.

Initially surveyed by parish or local area, the discussion in this book presents a synthesis of the results across the whole Gorge. The data were analysed first by considering the distribution of settlement and its characteristic forms. The study looked at the spatial relationships between buildings and industry as evidence for the organisation and control of land use, using the chronology of settlement to help map out the major phases of industrial development. The forms of settlement provide evidence for the changing structure and organisation of the industrial economy and society, as small and isolated settlements were gradually drawn together in an urban framework.

The study also considered patterns of development by looking at the relationship between buildings, and charting the distribution of single dwellings, pairs and terraces. These relationships provide direct evidence of the organisation of development and the processes which shaped the landscape of building. Thus the pattern of building associated with industrialisation began as a scatter of single buildings but soon encompassed the development of coherent terraces. This pattern of change came about through the piecemeal expansion of buildings as well as the deliberate building of groups, and is indicative of the location of resources in the economy and the emergence of a property market.

The codification of design and the architectural development of building also reflected the course of industrialisation: as a commitment of resources, buildings give at least some indication of a distribution of wealth. The large numbers of houses built by professional craftsmen for a population that was not engaged in agriculture in the early eighteenth century suggest the extent of a prosperity based on mining and manufacture; the gradual diversification of housing types suggests the emergence of a distinctive social hierarchy, especially during the nineteenth century. In housing, as in industry, development was constantly shaped by the inheritance of the existing building stock, and new architectural forms were themselves largely derived from existing models.

A comparative and quantitative study of buildings in the area identified the design processes used in building, and used this to classify the major local building types. There are two principal design processes for building: one relies on pattern books or the role of an architect, the other, which is more traditional or vernacular, appeals to and modifies the design embodied in existing buildings. Each process of design must work with a given development type and architectural vocabulary which comprises all the elements of building – constructional technique, plan, sectional form, detail and decoration. The role of traditional methods of design in the development of the Gorge is underscored by the identification of a common form and configuration of elements in many buildings, even where size and number of units in the plan varies. These similarities suggest that such buildings were built with reference to a common model. In other buildings, by contrast, the disposition of elements appears unique.

Conventional design methods could produce widely different types of building at different times: while the eighteenth-century cottage has much in common with the small classical house of the mid-nineteenth century, the access to a new architectural vocabulary and changes in the size and scale of new building mark the later building out. It is the combination of a process of design with a particular architectural vocabulary which defines the building type.

These types are historically specific – they are not just plan-type families or stylistic categories but belong to particular phases in the development of the area. They are of interest not just as a means of quantifying and phasing the development of industry but in so far as they give direct evidence for the development and structure of industrial settlement and society in which social stratification could be expressed even in hierarchies of architectural form and detail. The method of design and the architectural vocabulary used represent a series of choices available to builder and client, and reflect a social structure; how the building was to be used and who commissioned it are visible in its final form.

A building also represents a series of choices and requirements governing size and cost. These requirements will dictate what design and development process is used and will to some extent shape the choice of architectural vocabulary. The historical choices can be reconstructed by analysis of design process, development type and architectural vocabulary to enable building to be used as an index of social and economic change.[8]

Change and alteration were also considered as an important source for the history of the area, charting changes in land use and valuation and in housing standards. The study aimed not just to consider new building types and the sequence of architectural innovation but to look at the diversification of the building stock. The survival or modification of early buildings alongside the development of new forms was a key factor in the changing structure of settlement.

The structure of the study

The Ironbridge Gorge has long been the focus of historical research. Since the pioneering work on the technical and historical aspects of the Coalbrookdale Company Ironworks, the field has broadened to include ceramic and other industries. Historical illustrations and descriptions of the area have been drawn together in different volumes, and individual monuments have been the subject of many publications. Thorough synthesis of the area was presented in Barrie Trinder's *The Industrial Revolution in Shropshire.*[9] This body of historical research has cast light on some of the best-known historical remains in the area, particularly the Old Furnace in Coalbrookdale, and the Iron Bridge itself, and has certainly helped to ensure their preservation. In most cases, however, the starting-point has been the documentary evidence.

In presenting the results of the Nuffield Survey, it was decided to reverse this strategy and start with the physical evidence. This book therefore takes as its basis the surviving or documented field remains of the Ironbridge Gorge, and moves from there into historical analysis. Each of the following chapters, with the exception of the first, which sets out the general historical background, deals with a different class of evidence or technique of analysis.

The prosperity of the Gorge has often been attributed to its natural resources – the minerals, coal, ironstone, clay and limestone. But the way in which these resources were used was far from straightforward. Exploitation was not just a matter of availability but depended on a whole range of complex factors. In order to trace the relationship of mineral resources to the process of industrialisation, it is not enough just to deal with the natural landscape; the overlying pattern of human agency has to be deciphered as well (chapter 3).

The landscape could be as much a hindrance as a resource. Many of what at first appear to be the natural advantages of the area actually came to hamper industrialists, so the height which provided water-power inevitably made transport difficult, and declining navigation made the River Severn an awkward route in later years. But such adversity generated ingenious solutions, and some of the best-known developments in canal and railway technology were a direct result of the difficulty of negotiating the natural topography of the Gorge. One of the factors in the process of industrialisation is this relationship between the natural landscape and human activity, which is best deciphered through field evidence (chapter 4).

Industrial buildings are only just beginning to be seen as worthy of detailed architectural analysis, but are rarely the subject of archaeological attention. Even less is known about the slightly nondescript, very generalised buildings which are not specific to any particular process. It is these buildings which are least often protected. Yet those which do remain in the Gorge, mainly associated with the iron industry, highlight the emergence of a type of industrial vernacular in the form of the single-storey shed. This tradition survives today in the modern sheds of Telford, and is obviously worth further investigation.

Such buildings can also document the process of industrialisation. The study of the location and development of industrial buildings through time provides a stimulating perspective on the whole process of industrialisation. Many innovations have been claimed for the Ironbridge Gorge, and technical innovations are often held to be markers of industrial progress. And yet if the whole sample of industrial buildings which survives is considered, innovation appears to be the exception rather than the rule – it becomes clear that 'make do and mend' was far more common than the more celebrated periods of innovation. Thus, the study of the location and development of industrial buildings through time provided a different perspective on the process of industrialisation than that given in more traditional accounts (chapter 5).

The spatial distribution and phasing of house-building, the forms and patterns of its development, and the characteristics of its design can also evince the economic, social and cultural impact of industrialisation in the Gorge. These form the subjects of the second half of the book, which is concerned with how industrialisation happened in the spheres of domestic architecture, of settlement forms and planning.

Changes in the structure of settlement take place in the context of social and economic change, and can be used to chart the formation of an industrialised economy from its proto-industrial origins. Just as industrial buildings graphically demonstrate the tenacity of tradition, and the importance of adaptation rather than innovation, so in housing, too, each generation accommodated its inheritance of building, extending, modifying, but only rarely starting afresh. Nevertheless, a series of incremental changes throughout a long history of adaptation achieved a thorough transformation in the structure of settlement in the area, and an early pattern of scattered small hamlets became the framework for an urban landscape in the nineteenth century. Current management strategies which seek to conserve as much of this historical landscape as possible are only the latest of many revaluations of the built resources of the Gorge.

If the resultant pattern of settlement in the area looks at first sight haphazard, it can in fact provide a sensitive index of the organisation and process of industrial change. Housing reveals such a history through its location in relation to other forms of land use, through the relationship between its design and its use, but also through the history of its development. While development in the Gorge was *ad hoc* and individual in the earliest days of its industrial exploitation, the conditions were soon established for a market in land and housing, encouraging more coherent organisation, and both intensive and extensive land development. But this never amounted to extensive redevelopment, as land was released for building only in small plots, and its ownership was never consolidated in a few hands. Early prosperity endowed early buildings with a value for subsequent generations, and resources were never sufficiently concentrated to encourage wholesale replanning. As a result, the Gorge is a peculiarly rich historical landscape, each generation working within the constraints of its inheritance (chapters 6 and 7).

Such early prosperity is decisively illustrated in the early emergence of an industrial vernacular, for which informal rules of design and traditions of construction had evidently been formulated by the early years of the eighteenth century. The adoption of new vocabularies, the reinterpretation of existing models of practice, and the pattern of alteration can all chart the changing structure of an industrial society, with a minutely graded social and economic hierarchy carefully mirrored in the distribution of its building types (chapter 8).

The purpose of this book is to demonstrate the use, as well as the limitations, of physical evidence in studying the process of industrialisation in the Ironbridge Gorge. It is hoped that the results will have a practical application in the future management of this landscape, if only by setting out more clearly its historical structure. There is much more to be learned from the Gorge, and it is vital that future changes respect, rather than obscure, its history. Perhaps the principles underlying this study can also be applied to other industrial areas before their essential character, and the information they contain, is lost.

2

THE INDUSTRIALISATION OF THE GORGE

Introduction

Because the Ironbridge Gorge has been the subject of a great deal of historical research, it provides an ideal location in which to explore the relative contribution of landscape evidence to understanding the past. As early as the eighteenth century, visitors left accounts of the industrial wonders of the Gorge, and artists celebrated the achievement of the Iron Bridge and the spectacle of industry in a sylvan setting. Later visitors were rather more sceptical, seeing the once great area looking run-down and derelict, but these doubts gradually turned to an awareness of the role the Gorge and its personalities played in the Industrial Revolution. The local historian, John Randall, writing in the second half of the nineteenth century, was the first to describe the breadth of industries in any detail. In the 1950s, as part of a growing national interest in industrial archaeology, there was a renewed interest in the Gorge, focused on the technical history of the iron industry. Such work was instrumental in the establishment of a small museum at the old ironworks, and eventually the formation of the Ironbridge Gorge Museums. The expansion of this group of museums to include a chinaworks and a tileworks generated further new work on the whole range of local industries.[1]

This chapter draws heavily on this previous research, seeking to set out the history of industrialisation from the medieval period to the present day. The aim is to present a wider context for the better-known events of the Industrial Revolution period. This provides a useful background to our exploration of the landscape evidence.

The medieval foundations of industry

There is very little evidence for occupation in the Gorge before the Norman Conquest. In this part of Shropshire, ancient woodland survived to a very late date and there was little settlement. However, there were at least two clearings in the vicinity of the

Gorge, one at Madeley and one at Caughley, estates which had been given to the double monastery at Much Wenlock, founded in 690.

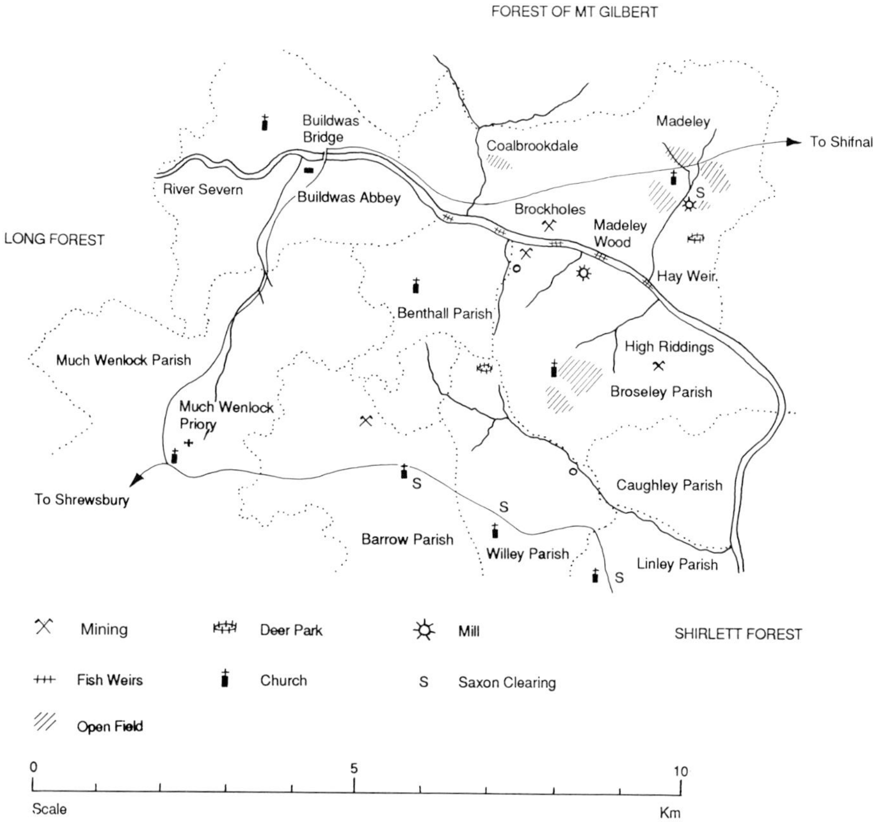

2.1 Medieval sites mentioned in the text (Shelley White)

The boundaries and lands of Wenlock Priory were the most important influence in setting the pattern of land use and ownership in the Gorge until the formation of Dawley New Town. The early monastery at Wenlock fell into disuse, but was refounded in the eleventh century with most of its lands intact. The Domesday survey shows that it included the manor of Madeley, the parish of Little Wenlock on the north side of the Severn, and the large parish of Wenlock on the south which included the area of the later parishes of Broseley, Benthall and adjoining parishes (fig. 2.1).[2] Much Wenlock remained the administrative centre for Dawley, Madeley, Broseley and Wenlock until 1966.

Shortly after the Conquest, much of south Shropshire was designated as a royal forest under laws which preserved land (not necessarily wooded) for hunting, and restricted settlement, agriculture and the exploitation of minerals. The parishes of the Gorge fell into the Long Forest, and thus the freedom of Wenlock Priory to develop its lands was proscribed. The monks were fined for clearing at Benthall in 1125 and at Broseley in 1250, while at the same time the Crown was able to raise money by granting private estates in the forest in Benthall and the eastern part of Broseley. With disafforestation, from the thirteenth century, the pace of clearing increased but much of the Gorge nevertheless remained wooded. Woodland remains on the slopes even today and much of it can be shown to be of ancient origin.[3]

A second abbey was founded in 1135 by the Savignacs (later Cistercians) at Buildwas. In contrast to Wenlock, little attempt was made to build up a local estate; instead the monks acquired scattered lands in Staffordshire, Derbyshire and around Chester, and farmed their own small local estates directly.[4] Thus, although the supply of local raw materials and produce was important to the abbey economy, Buildwas cannot be said to have had the same long-term impact on the local landscape as Wenlock.

The present pattern of settlement was largely derived from the policies of Wenlock Priory, which controlled the parishes of Madeley, Benthall and parts of Broseley until the Dissolution. In Broseley, the medieval centre around the present church was surrounded by open fields and patches of woodland, with common land to the west.[5] The east of the parish was dominated by large freehold farms, which perhaps began as forest grants. The priory kept control of a third of the parish, some farmed directly and other parts by tenants, and had another large estate in Benthall around the present church. Traces of the deserted village which once formed the centre of Benthall parish can be seen at the back of the hall. In Madeley the medieval settlement probably lay around the present church with an outlying settlement at Coalbrookdale where open fields were noted in the thirteenth century. When the priory was granted permission to hold a market in Madeley from 1269, a 'new' town was in effect created – about 60 burgage plots were laid out along the newly improved road from Shifnal to Wenlock, and a market-place was set in the centre.[6]

Small-scale industry was another source of income. By far the best-documented early industrial activity was the construction of mills primarily for grinding corn. The River Severn itself was too variable in height for use as a source of water-power, but the many small streams dropping down into the Gorge were ideal. There were mills on the Washbrook in Madeley, just to the north of the present Blists Hill site. Another mill lay in Broseley in 1188, there was one at Benthall in 1317, and two by 1545 in Broseley. The sixteenth-century mill at Coalbrookdale was almost certainly not the first mill in the valley. The impact of these early water-powered installations is apparent in the landscape. Although the mills no longer survive, the valley pools created to supply them with water remained a lasting landscape resource, to be reused by later industrialists (see Chapter 4). Another landscape survival is the medieval fish weirs on the Severn, which remain obstacles to navigation even today.[7]

Building materials for both Wenlock Priory and Buildwas Abbey were found in the Gorge. Buildwas lies in an alluvial valley with little local stone. Limestone for mortar and for walls came from Benthall Edge and sandstone for facings and decorative work came from Broseley by river. This was most likely the same source of the golden coal-measure sandstone as was used in the construction of Wenlock Priory (although the red sandstone would have come from Bridgnorth). Both were paved with decorative tiles, the clay for which, when fired, closely matches that from the nearest outcrops in Broseley and Jackfield.

In contrast with evidence from the Forest of Dean or the Weald, there is little to suggest that the Gorge was a substantial medieval mining area, even though local coal was low in sulphur and suitable for both domestic and industrial uses. There are scattered references to mining – coal was brought from Benthall to Buildwas in 1250, and mined near modern Ironbridge in 1322. Ironworking was equally limited: Wenlock had forges at Shirlett and an ironstone quarry at Little Wenlock, and the monks of Buildwas may have had a small bloomery. Strong monastic control had created a situation in which neither the free miners who had opened up the Forest of Dean nor the entrepreneurial landowners of the Weald were able to operate. Only the early mining at the High Riddings in Broseley just prior to the Dissolution suggests any activity by freeholders.

Thus the long-term effect of monastic involvement in the Gorge was the centralisation of what little industry there was under monastic control. Even in agriculture, direct farming had been replaced by the extraction of heavy rents as a source of income. Settlement in Barrow, Willey and perhaps Coalbrookdale contracted and the area remained wooded. The valuable mineral resources had barely been touched, except to supply the local needs of Wenlock and Buildwas. At the Dissolution, the Gorge presented an area of great opportunity.

Land-ownership and the origins of industrialisation

On 26 January 1540, Wenlock Priory was dissolved and its lands broken up and sold (fig. 2.2). Within a century, the Gorge was a coal-mining area of national importance and new industrial settlements had spread over the old agricultural commons. This change in the scale and pace of industrialisation was the result of the activities of four men: Sir Basil Brooke, John Weld, Lawrence Benthall and James Clifford. They (or their families) had purchased former monastic lands from the Crown to add to existing inherited estates. Fully aware of the potential of these lands, they set about utilising neglected resources. With capital from other interests, and control of minerals, timber, transport and often labour on their own land, they were in a unique position to exploit the area to the full.[8]

In Broseley, James Clifford had inherited two-thirds of the manor, and was able to acquire the remaining third which formerly belonged to the priory. He leased out the farming lands and set about exploiting the minerals himself. Through the efforts of Clifford and others, the coal industry transformed from one supplying the local, domestic market, to an export-orientated one, supplying a national demand for coal via the River Severn. Adits were dug directly into the coal seams from the hillsides, and in order to get the bulky mineral down to waiting barges on the river Clifford constructed a sophisticated transport system with early inclines, and some of the first wooden railways in the country.[9]

John Weld, a London lawyer who purchased an estate at Willey, was even more

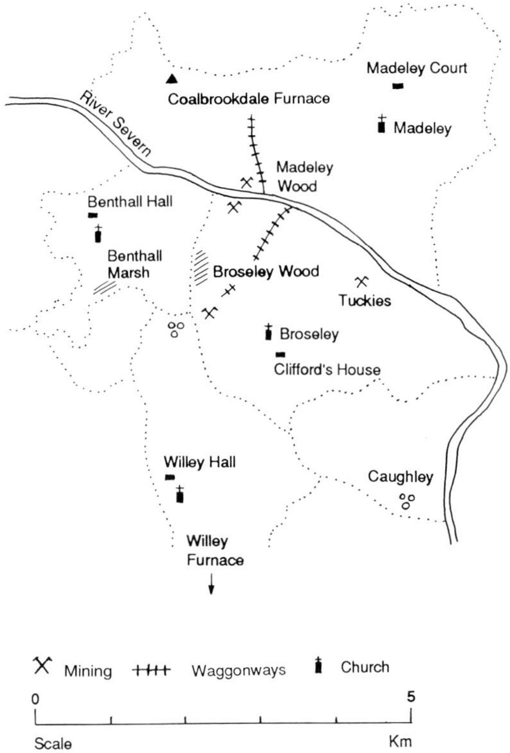

2.2 Post-dissolution sites mentioned in the text (Shelley White)

intent on maximising returns from all of his properties. As well as enclosing lands and making agricultural improvements, he opened coal-mines, using his own timber for props and construction. Most of his estate lay away from the river, and in order to get his coal to market he was forced to negotiate with his neighbours to lay wagonways across their estates. Weld spent £500 on an old charcoal blast furnace at the Smithies, and was able to smelt his own iron.[10]

By 1645, the township of Benthall was one of the largest sources of coal on the Severn, producing 30,000 tons of coal a year, yet at the Dissolution it had been only a tiny, wooded parish. Lawrence Benthall II, the grandson of one of the post-Dissolution speculators, presided over the transformation of the parish. By the 1630s, he had dug mines, laid railways, built cottages and imported labourers.

Mines in Madeley parish had first been opened up by John Brooke. His son, Basil Brooke, had widespread interests in coal and ironworks in the Forest of Dean. He used capital from these enterprises to expand coal-mining in the parish, and to operate an ironworks. In the 1620s he was experimenting with making steel, very possibly at the old timber-framed malthouse (now demolished) in Coalbrookdale, which was described in 1734 as a former steelhouse.[11]

Mining and ironworking required labour, and in this traditionally agricultural area of closed communities labour was in short supply. Clifford, Weld and Benthall built cottages for their workmen and encouraged them to settle on the old commons. In Broseley Wood, for example, a classic 'squatter' pattern of settlement was created with houses dispersed on small random plots. The so-called 'scummes and dregges of many countries'[12] who came into the area caused the older agricultural tenants to riot over their lost common lands.

But the landowners may have had another motive in settling people on commons. Most of the Broseley and Benthall commons were on marginal land that was poorly drained or on steep slopes with poor soils. It was on precisely these types of land that coal was likely to occur – outcropping on the valley sides, or lying just under the surface of marshy areas. There was therefore a direct association in these parishes between commons and coal, and rights over this coal were of intense interest. By settling their own tenants on common land, the landowners were attempting to establish their right to mine that land. When Benthall and Weld entered into a dispute about old grazing rights on Benthall Marsh, their underlying argument could well have been about the right to mine and to lay rails, and when Benthall built cottages in Broseley Wood he could equally have been claiming a right to mine there.[13]

In the period leading up to the Civil War, there is evidence for industry on another scale, the work of a class of industrialists with less capital to spend. Shallow coal-mines, with their characteristic ring of spoil (known as bell pits) are to be found scattered through the woods of Broseley, Benthall and Caughley. These pits are difficult to date, but were probably dug by individual miners in the seventeenth century without the capital to pump mines or build railways on any large scale. These miners could have cited manorial rights in order to stake their claims to underlying coal. Other individuals set up small forges and ironworks; pieces of charcoal-rich iron slag can be seen in the Caughley woods and along Linley Brook, and there was an old smithy elsewhere in Willey. There were local brickmakers supplying Weld and Benthall with bricks, and the beginnings of a clay pipe industry in Much Wenlock and Broseley.

Coal and industrial diversity

The Civil War saw the end of the monopoly of the landowners in local industry. Weld, Brooke and Benthall were all Royalists who were fined heavily and had their lands confiscated. The Coalbrookdale ironworks were sequestrated, and the Madeley mines fell into disrepair. The coalworks of Benthall had been of such strategic importance to Shrewsbury and Bridgnorth that the hall was garrisoned, but it later fell to the Parliamentarians.[14] Coincidentally, much of the easily accessible coal had been worked out, with the result that production was never established again on the same relative scale.[15] Clifford had died earlier in 1620, and all his Broseley lands were dispersed in small lots to pay off creditors. For all their efforts, investment in industry had generated no lasting prosperity for the landowners. Weld advised his successors not to be 'busy in searching for coals, nor in Ironworks', and turned increasingly to agriculture.[16] The Benthall estate became an agricultural one, and Broseley and eventually Madeley were broken up into smaller units. Yet by building roads and tramways, digging mines and encouraging new industrial settlements these landowners had created a solid industrial infrastructure within the landscape which was to become the base for the next phase of industrialisation.

The break-up of the landed estates created an opportunity for individuals with less capital, and often no freehold land of their own, to take the initiative in coal-mining and industry. This was most apparent in mining, where probate inventories document the emergence of the charter master, or master collier, who took over responsibility from the landowner for getting coal, laying railways and maintaining mines.[17] Such men were wealthy members of society, and built themselves timber-framed houses on the riverside, away from the traditional agricultural lands. Several of these houses survived into this century.

These industrialists used their money to improve their mines and transport systems. New and deeper mines were dug in Broseley Wood and on the high ground in Madeley. To service these mines, horse gins were used for winding and pumping, and in 1719 a consortium of miners installed the earliest steam engine in Shropshire to pump water from a shaft which can still be seen in Madeley Wood.[18] The transport network linking these mines with the river was also improved. New inclines or 'winds' for letting coal down to the riverside were built, and mining leases often included the right to lay rails for wagonways. As a contemporary observer remarked, many of these routes were short-lived and taken up at the expiry of a lease. Few of the documented routes of this date are actually visible on the ground, but by the early eighteenth century there is evidence for more permanent routes, built level with embankments and made-up surfaces.[19]

Associated with the development of transport from the mines came the improvement of the riverside itself. Weld and Clifford had wharfs for loading their coal on the river bank, which were probably little more than enclosures. Clifford was also mining close to the riverside – he was accused of dumping spoil in the Severn from mines near the Tuckies in Jackfield. By the mid-eighteenth century, permanent wharfs were built at the end of key tramway routes such as the Meadow and Loadcroft wharfs on the north bank, and Lloyd Head and the Tuckies on the south bank. The river, however, was never improved. Despite persistent pressure to improve navigation by the construction of locks, there was strong local opposition, and the river level continued to fluctuate wildly from season to season. The banks constantly eroded, old fish weirs such as 'Eve's Mount' impeded traffic, and for many months of the year navigation was impossible.[20]

The evidence for riverside developments underlines the persistent links between pits and the River Severn, and implies a continuing export trade in coal. But at the same time new uses for coal attracted a group of diverse industries to the Gorge, all of which can in one way on another be linked to the availability of local coal supplies.

The most noticeable industrial feature to an observer standing near the river in 1746 was the lead smelter at Bower Yard, 'vastly poisonous and destructive to

2.3 The Benthall lead smelter may be seen to the right in this view of the Iron Bridge
Francis Chesham (after George Robertson), 1788 (IGMT AE 185. 765)

everything near it' (fig. 2.3).[21] There were three or four such smelters in the eighteenth century, all located on the riverside and using ore from Wales and coal from local pits. Lead smelting was one industry for which there seems to have been little rationale in the Gorge apart from abundant coal supplies and perhaps a certain tolerance for noisome and polluting industrial enterprises. Another industry directly associated with coal-mining was the extraction of mineral tar from local sandstone by Martin Eele at the Calcutts. The tar was used to caulk boats, or distilled to produce varnish and medicinal oils. Just to the east were the Salthouses, where natural brine was collected from coal pits and boiled to produce salt.[22]

Tar manufacture and salt-making were relatively minor operations in comparison to the two main coal-related industries of the area – ceramics and iron. By the late seventeenth century, the manufacture of pottery, clay tobacco-pipes, bricks and roof-tiles was well established using the variety of clays which came as a by-product of coal-mining. Most of these enterprises supplied a local market, generated by the growing industrial settlements.

Two distinct pottery-making traditions persisted in the Gorge – the manufacture of coarse earthenwares for local use, and the production of a small amount of fine table-wares for an export market. Fine white salt-glazed stoneware was made at Jackfield and in Broseley from the early eighteenth century, in coal-fired kilns. Typical salt-glaze saggars are to be found in garden walls throughout these areas. This industry was short-lived, but almost every garden on the south of the river produces sherds of yellow and brown slip-trailed pottery and coarse, dark earthenwares. By the mid-eighteenth century there were at least four places at Jackfield and another at Benthall making such wares in domestic workshops.[23] Clay tobacco-pipe manufacture also operated on a small scale from the 1640s. Using fine white clays from local coal measures, pipes were made as a sideline to other activities, and distributed widely through Wales, Staffordshire, Cheshire and the West Midlands.[24]

Buildings, better than any other source, document the growth of the brick industry. Prior to 1700 brick was a rarity, used on the occasional large house such as that bought by Weld at Willey, but from this date onwards the use of brick in domestic buildings increased dramatically. A great number of small cottages were built to house the growing population of coal-miners, bargemen and labourers, all of similar form, and all built in brick (discussed further in chapter 8 below). Industrialisation came about not only because of the availability of coal, but through the growth of settlement associated with mining.

Abraham Darby's use of coal in metal-working and iron-smelting is perhaps the best-known example of the role of local coal in industry. When Darby arrived in Coalbrookdale in 1709, he found a charcoal blast furnace that had been out of operation since 1705 when the dam had burst and his predecessor, Shadrach Fox, had absconded. Darby leased forges, charcoal houses, a steelhouse and other buildings, all linked together by a complex water-power system which may have dated back to the sixteenth century.[25] There were a few cottages, but not the populous mining communities of other parts of the Gorge. Darby established a complex pattern of manufacturing, apparently intending to produce iron and copper, and make brass at Coalbrookdale, and to work brass and iron at Tern.[26] For the first time, Darby was able to use coal rather than charcoal to smelt iron. Unfortunately, the iron he produced was suitable only for casting and not for making wrought-iron, the most versatile form of iron. The market for his iron was thus a limited one. Darby's operation therefore remained small – he and his employees sold the pots at markets and fairs in exchange for cash in the same way as many other traders (fig. 2.4). It was only after Darby's death that the operation began to expand, and to deal directly with other industrialists.

The success of all of these industries depended not only on coal but also on labour. There were two factors which encouraged the growth of an industrial population: the availability of land for settlement, and the opportunity for a full seasonal round of employment. New communities of miners had been established earlier on the old commons of Madeley and Broseley, but by the mid-eighteenth century they had trebled in size. Settlements which had begun as casual, unregulated scatters of housing had become almost prosperous suburbs. It is clear that these were not agricultural households – they were located well away from farm land, the plots in which the houses stood were small and very few occupants owned cattle or agricultural implements at their death.[27]

Yet industry was unlikely to have provided year-round employment. Coal-mining was largely concentrated in the summer months when the water-table was lower and pumping less of a problem. Coal could only be transported by river in winter when

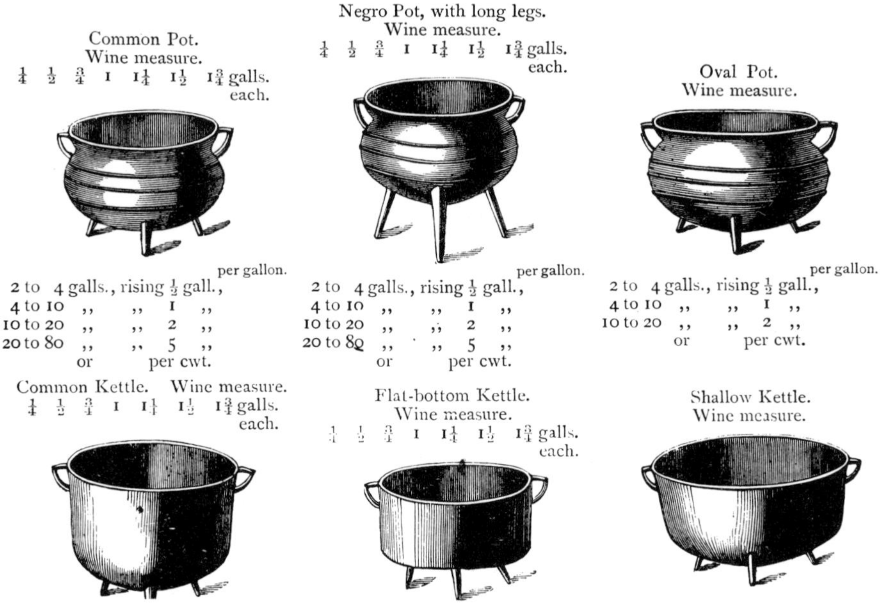

2.4 Cast-iron pots advertised in the Coalbrookdale Company catalogue of 1875 (IGMT 1989.4311)

the levels were high enough for barge traffic. Clay mined in the summer months was weathered over the winter in readiness for building work during summer, so brick and pottery production took place largely in the spring. Iron production was concentrated in the autumn and winter, rather because there were shortages of water to power the bellows in the summer. The seasonal nature of many jobs allowed industrial work to become a more viable occupation. There was also scope for part-time involvement in low-capital manufacturing such as clay pipe making, or pottery-making in a kiln attached to a row of cottages.

A common model for the origins of the Industrial Revolution is based upon rural communities supplementing agricultural income with part-time industrial work such as spinning or weaving. This form of production has been termed 'proto-industrialisation' and was typically on a domestic scale, required little capital, and was geared towards national markets. Proto-industrialisation has been seen as a first step towards the large-scale industrialisation which took place in, for example, the textile industries of the late eighteenth century.[28]

Instead of an agricultural population, there was in the Gorge an industrialised population, turning to by-employment to supplement a seasonal income. The importance of this pattern of work was that it enabled an industrial labour force to grow and survive. The ready availability of coal had encouraged a diverse range of industries which could support a full-time industrial labour force.

In the period up to 1750, there had emerged a landscape which was to provide the base for an extraordinary expansion in local industry. It was not so much the availability of a single raw material – coal – but the diversity of natural and man-made resources it generated, which was to enable industry to expand so rapidly. Those who had made fortunes from industry in the earlier eighteenth century grouped together into partnerships which had the capital to integrate what had hitherto been different industrial activities.[29] These new, powerful, co-ordinated operations were the ironworks and factories of the later eighteenth century.

The Industrial Revolution in the Gorge: integration, innovation and the factory system

To what extent can an industrial revolution be recognised in the Ironbridge Gorge? Certainly between 1745 and 1805 a very real transformation took place in the industrial landscape. There was a phenomenal growth in ironworking as five vast new works were set up in the Gorge. The construction of tramways, canals and roads brought better links with the rest of Britain. Three new porcelain factories were set up, employing hundreds of workmen in purpose-built factories. Even the long-established brickworks and potteries increased their scale of production. Behind this transformation lay a new form of industrial organisation: integration. Entrepreneurs and small partnerships of colliers had evolved into integrated companies which were able to control more of the means and processes of production than their predecessors. Raw materials, production, transport and marketing were all integrated into a single operation. With more capital they were able to build bigger works,[30] to use the newest technology and to produce for a wide market, beyond the Gorge.

Although the iron industry dominates accounts of the Gorge in the late eighteenth century, it was in the ceramics industry at sites like Caughley and Coalport that the greatest changes took place. In the late eighteenth century, the introduction of tea-, coffee- and chocolate-drinking had created a market for fine teawares imported from China. In response to this growing demand, European manufacturers experimented with the production of porcelain in imitation of the Chinese blue and white wares. At Caughley, in the fields to the south-east of the Gorge, Thomas Turner of Worcester built a factory employing about 100 people to manufacture the new soft-paste porcelain. Fine clay and flint were imported by river and brought probably by rail to the site. Local coal, clay and fireclay were used for saggars, to fire the kilns and for the new buildings. At first wares were sent back to Worcester for decoration, but later blue printing was done at Caughley itself and the finished products dispatched by river. On what today seems a greenfield site, Thomas Turner was able to utilise coal-mines, tramways leading to wharfs, a saggar works and an earlier earthenware works. He was able to integrate these existing resources with his own network of contacts in Worcester, to set up a business which was far removed in size and complexity from the earlier domestic workshops of Jackfield.[31]

Caughley remained an isolated phenomenon until 1796, when a second large factory was built on the riverside at Coalport to manufacture creamware, a fine earthenware developed in the 1740s. Creamware was marketed assiduously by Josiah

Wedgwood and became one of the first British wares to be exported on a large scale. At Coalport, Walter Bradley made a range of other wares in addition to creamware, but seems to have operated for only a few years. Much more successful in the long term was the factory built nearby by a former employee of Thomas Turner, Edward Blakeway. With two other partners, he built a new porcelain manufactory opposite Bradley's works. A third factory was set up to make porcelain in 1800, on the site of (or reusing) Bradley's works, but operated only briefly until John Rose took over the whole Coalport site in 1814, becoming the sole manufacturer. Together, these works employed hundreds of people in a range of three-storey buildings. Many different specialised processes were integrated into one operation – grinding raw materials, making saggars, producing and decorating wares, and eventually marketing, as the company acquired outlets locally and in London.[32]

By choosing Coalport as a site, these works were taking advantage of a new link between the river and the canal system established by William Reynolds. Reynolds's father had purchased 20 acres of land at the Sheepwash in 1787/8, a 'rugged uncultivated bank which scarcely produced even grass'.[33] With the completion of the Shropshire Union Canal to the top of the hill above Coalport, an inclined plane down to the riverside and a length of canal parallel to the river, the site became a prime industrial location.[34] William Reynolds constructed a massive sandstone quay, and a large warehouse spanning the canal and river for transferring goods from one to the other. He saw the area as a potential industrial estate, with different industries 'united like links in a chain', and encouraged a timber-yard, ropeworks, chainworks, a brickworks and a bag factory to be set up. In partnership with Lord Dundonald, he planned a huge chemical works on the site, but it was never built. Although Coalport was Shropshire's most important river port by 1797, it never became the new town Reynolds had envisaged. Only the chinaworks, the timber-yard and chainworks operated for any length of time and, despite Reynolds's plans for housing, most workers continued to live on the south bank.

In the late eighteenth century, another transformation occurred in the way common earthenware was manufactured. There was a new pottery at Haybrook in 1743, another at Caughley by 1754, and by 1773 there were three factories at Jackfield and a new works at Benthall. There were further potteries at Ladywood, Coalmoor, on the top of Bridge Bank in Benthall and at Horsehay.[35] All of these were located on the buff clay outcrops, near coal-mines, and often on main turnpike roads away from the river so that pottery could more easily be sent to market. But the key to this new scale of operation seems to be that several, at least, were making slipwares using new mass-production techniques. Early-eighteenth-century wares are coarse, handmade pieces with decoration trailed on in thick lines. Yet the fields around Broseley are strewn with a standardised type of pottery, flat with a pie-crust edge and very thin yellow and brown slip trailing.[36] The press-moulding technique for making these had been known since the seventeenth century.[37] However, in the eighteenth century, plaster moulds were introduced, enabling this pottery to be mass-produced to a standard design. Slip and powdered glaze were then applied to the wet clay before it was fired in a coal-fired kiln. Plaster moulds and single firing made the wares quick and easy to produce, and relatively cheap (see Chapter 5).

In the iron industry a similar transformation took place.[38] Coke-smelted iron had been made in the Dale for almost fifty years, in furnaces no larger than the average charcoal furnace. Yet in the space of a decade in the 1750s four new works were built in the Gorge at Madeley Wood, Horsehay, Calcutts and New Willey to the south. These new works had larger furnaces, produced more iron and employed more people, but the most obvious way in which they differed from earlier works was the degree to which operations were integrated. Nothing makes this new conception of an ironworks clearer than the grand design for Newdale, a purpose-built works set up to the north of the Gorge. There, on a greenfield site, houses, ironworking buildings and perhaps a blast furnace were laid out; a tramway was built to nearby mines and limeworks, so that the full range of ironworking activities were organised to take place at one location.[39]

Mining was the first activity to be integrated with ironworking. Abraham Darby had been supplied with raw materials by individual contractors,[40] but from the 1750s his successors, the Coalbrookdale Company, began to lease their own minerals at Lawley and Dawley and to take responsibility for supplying their own coal and ironstone. This was taken a step further when the company bought up land in the the Manor of Madeley and leased the mineral rights to most of the parish.[41] And as ironmasters turned to coal-mining, so miners turned to ironworking. The Madeley Wood Company was originally set up to mine coal, but soon built iron furnaces on the riverside at Bedlam, and a partnership of colliers built another new ironworks at Lightmoor. The coal industry was no longer totally orientated towards export, and iron and coal-mining had become, perforce, an integrated operation.

Ironworking as an activity also began to move north, as an increasingly large percentage of the coal and ironstone for the works was obtained from the deeper seams in the northern part of the coalfield. New works were built at Ketley and Horsehay by the Coalbrookdale Company. This was not the only large organisation, however. A classic extension of the trend towards integration can be seen in the development of the parish of Lilleshall. Here coal had been exploited on a piecemeal basis since medieval times by charter masters, paying royalties to the lords of the manor. As the demand for minerals increased, a new partnership was set up by the landowners to take over mining and to invest in the area – a canal was built to carry minerals across the estate, and later blast furnaces were built at Snedshill (1780) and Donnington Wood (1785).[42]

A second group of ironworks built on the south bank of the river displays a rather different rationale. Furnaces were built at Barnetts Leasow, Calcutts, Benthall, Broseley and Coneybury. Part of the reason for this was that forging remained an activity separate from iron-smelting, still dependent on charcoal fuel and water-power. An eighteenth-century forge still comprised a finery, a chafery and a hammer – a technology little changed since the medieval period (fig. 2.5). Despite the development of steam-power, even in 1800 forging was 'chiefly carried out by means of trifling brooks', according to Thomas Telford. In 1773 a new technique for producing

2.5 Forging iron in a small hand-operated forge

wrought-iron with charcoal was patented by John Wright and Richard Jesson. Small pieces of iron were heated in clay pots and thus did not come into contact with the coke used as fuel. Known as 'stamping and potting', or the Shropshire process, this method was adopted by the ironworks at Horsehay, Old Park and Ketley.[43] But new forges continued to be located on the River Severn away from the main ironworking area – at Wren's Nest, Eardington and Hampton Loade. It was partly to supply this river-based demand for forging iron that the Barnetts Leasow furnaces were set up on the south bank of the river in 1796.

The other aspect of the south-bank industry was casting. Specialist castings had been produced at Coalbrookdale for many years, but the widest demand was for domestic and agricultural items cast in the foundry. Foundries at Benthall and Broseley ironworks were able to take advantage of cheap local supplies of iron but otherwise were no different from numerous other foundries in towns throughout Britain. These works had access to road as well as river transport and were essentially supplying a local consumer market, unlike the specialist works to the north.

As a strategy, the integration of an existing industrial landscape had allowed industrialists to build and operate bigger and more profitable works than had hitherto been possible. But with the rapid expansion of industry elsewhere in the country, the ever larger scale of operations and generally falling prices in the early nineteenth century, integration was no longer a profitable option. Despite new canals and turnpike roads, the Gorge remained, as it always had been, an isolated area, away from the main centres of industry and population. With its high transport costs and by now outdated working practices, production in the Gorge was becoming increasingly uncompetitive. In the mid-nineteenth century, producers reversed their strategy and switched from integration to specialisation.

Specialisation and the production of prestige goods

By the early nineteenth century, iron production in the Gorge was outmoded and uneconomic. Matters reached a nadir between 1815 and 1822 when there was a depression in the price of pig-iron through the ending of the Napoleonic wars and the availability once more of continental supplies to Britain. All five of the Broseley furnaces were out of production in the early 1830s and iron-founding continued for only a short time afterwards. Part of this collapse was due to a shortage of local raw materials,[44] but the extraction of ironstone continued long after smelting ceased. The Gorge was fundamentally an isolated location, too far from markets for either iron or raw materials. Even the opening of the Shropshire Union Canal wharf at Wappenshall in 1835, which linked the Gorge for the first time with the national canal system, brought few benefits to the south bank.

The fortunes of the Coalbrookdale Company were somewhat different. In the early years of the century the works were in a poor state – the furnaces were out of blast and the great steam engine which recycled water was being demolished.[45] One of the Horsehay furnaces was closed, the Ketley works ceased operations in 1816 and the possibility of closing Coalbrookdale altogether was raised. The situation was similar throughout the coalfield. Unlike the works on the south bank, the Coalbrookdale Company was able to adapt to these changes. Following a general improvement in the iron market and a change of management in 1830, the Coalbrookdale Company enterprises were reorganised. The forges at Coalbrookdale were closed and the Dale specialised in casting and foundry work, particularly pumping engines and domestic castings. Forging was moved to Horsehay, where new puddling-furnaces and rolling-mills were built, and pig-iron was supplied from the newly purchased Lightmoor furnaces. No longer did one works undertake the full range of ironworking activity.

Specialisation was more than just a matter of reducing the range of activities; it involved the production of a limited range of prestigious goods aimed at a volatile consumer market. In this way, producers could add value to their goods to overcome the problems of transport out of the Gorge. At Coalbrookdale the company became renowned for art castings (fig. 2.6). Under Francis Darby and Abraham Darby IV, fountains, sculptures, gates and garden furniture were produced to the designs of artists employed by the company and later encouraged by the setting up of an art school in the Literary and Scientific Institute in 1859. Through its display at the Great Exhibition of 1851 the company became internationally famous for such

2.6 An elaborate cast-iron fountain taken from the Coalbrookdale Company Catalogue of 1875 (IGMT1989.4313)

products, and exported them around the empire. Medals were awarded at various other exhibitions, and in its publicity the company claimed to be the biggest ironworks in the world. The company further attempted to enhance their reputation through the manufacture of terracotta at Lightmoor. Moulded statues, architectural faience, vases and garden products were added to the more mundane output of bricks, tiles and drains, but production was abandoned after only a short period.

In the northern part of the Coalbrookdale Coalfield, smelting continued, but although production was at its highest in 1869 this was only 2 per cent of the national output in comparison to 40 per cent in the 1770s. In absolute terms, the importance of the coalfield was declining. As new processes such as Bessemer steel-making were introduced, ironmasters preferred to undertake new ventures in more convenient locations rather than rebuild old and decaying works. The crisis in the coalfield was reached in the 1880s when all but two or three works closed down. The Horsehay site was sold, but a new partnership continued in operation as an engineering firm specialising in the assembly of bridges and other complex iron structures.[46] The strategy was so successful that production at the site only ceased in 1985.

During the early nineteenth century, there was an extraordinary growth in another very specialised industry – the manufacture of dark purple clay roof-tiles.[47] By 1840 the Tithe maps show that there were no fewer than thirteen brick- and tile-works in the Gorge and, of these thirteen, most were principally producing roof-tiles. This level of production may have been linked to the new and deeper coal-mines, which brought access to the best seams of clay for making tiles. What is so unusual about this industry, however, is the size of the works; contemporary brickworks elsewhere consisted of little more than a few thatched shelters and an open-air clamp or kiln, but at Jackfield there were well-organised works on a large scale all producing the same things.[48]

These factories must have been producing for a wide regional market. Output far exceeded the local demand for roof-tiles, even though almost every building in the Gorge seems to have been reroofed. The works were located on the riverside, away from the main roads or centres of population like Broseley and Madeley. Goods must have been exported by river and perhaps canal to a wider market. Nevertheless, the construction of the Severn Valley Railway in 1862 was a boon, and several works built their own sidings in order to dispatch goods directly by rail. The railway created a demand for tiles which existing processes could not meet, and new mechanised production techniques were introduced. At the same time, the railway network made Welsh slate more competitive, and the roof-tile works came under increasing pressure to produce cheaper goods. Unfortunately, when they did so, the quality decreased and Broseley tiles acquired a poor reputation. Production exceeded demand, and because they were laid out for specialist tile-making they were unable to diversify into brickmaking. By 1940 almost all the works had closed – labour shortages, transport costs, clay shortages and outdated works all contributed to make them uneconomic. Broseley tiles did not play a part in Britain's post-war building boom.

This trend towards low-bulk, specialist goods for a national market was nowhere more obvious than in the encaustic tile industry.[49] By 1883 Jackfield was the location of two major manufactories with an international reputation for decorative tiles. The tiles of Maws and Craven Dunnill were to be found in houses, churches and public buildings around Britain, and were some of the few products of the Gorge to reach the outposts of the empire. Plain red floor-tiles had been made in the eighteenth century, but the earliest attempts to produce more elaborate tiles took place at a Jackfield works in the 1830s. Encaustic floor-tile production was initiated on a more permanent basis by the brothers George and Arthur Maw at Benthall on a site fashioned out of the old Benthall ironworks. They were principally attracted by the good clays of the Gorge, which had already been transported down the river to their Worcester site.

Production continued at Benthall for almost 30 years, during which time many new processes for making tiles were patented. But the site proved to be crowded and inefficient, and so the company moved to a new $5\frac{1}{2}$-acre site on the riverside to the east of Jackfield. Here in 1882 they laid out a complete purpose-built works designed by the architect Charles Lynam of Stoke-on-Trent, to incorporate all the production processes in a logical whole. But before this factory was built Charles Lynam had designed another encaustic tile factory for the Craven Dunnill partnership. This partnership took over an old pottery which had recently been used for making geometric tiles. The pottery was demolished and a new factory built which would facilitate the speed and efficiency with which clay could be moved around the site.

The works were very successful, initially capitalising on the demand for reproduction medieval floor-tiles created by the programme of new church building and restoration in the mid-nineteenth century. New and more elaborate finishes and glazes were developed, and tiles became increasingly ornate. Like the Coalbrookdale Company, Maws and Craven Dunnill both gained an international reputation for high-quality, specialist products.

The same trend towards prestigious products can be seen in the attempts to establish a local art pottery industry.[50] While the mass-production of common earthenwares continued at sites in Jackfield, at Benthall and on Bridge Bank, two potteries moved briefly into the production of art pottery. Classical urns with incised decoration and extraordinary pieces with moulded fruit and vegetables were produced for a short period. Although art pottery never thrived, the fact that expensive, individually produced decorative items were seen as a more successful business proposition than mass-produced domestic pieces reflects the situation of the Gorge where transport costs restricted large-scale production.

Despite the construction of new roads and canals in the eighteenth century, and even three different standard-gauge railways in the nineteenth century, transport remained the biggest barrier to commercial success in the Gorge. Raw materials were running out, and there were few large markets nearby. Manufacturers realised that quality and reputation were more important than mass-production and cheapness to the continuity of local industry, and specialised in the production of a limited range of high-quality finished goods. Yet it is debatable to what extent prestige goods were economic; plain tiles with simple repetitive motifs became the staple of Maws and Craven Dunnill, and at Coalbrookdale the market for domestic stoves and bricks became more important in the later nineteenth century than ornate statuary and terracotta urns. While companies acquired a reputation for prestige goods, their fortunes were based on much more mundane products.

Twentieth-century pioneers

At the outbreak of the Second World War, a visitor might have seen a dismal future for the Gorge. Changing fashions left no market for the prestigious iron and ceramic industries, and with the labour shortages of the war, ailing industries had closed for ever. Yet, ironically, it can be shown that industry continued to operate in often novel and pioneering forms throughout the first half of the twentieth century, and it was only through official policies in the 1960s and 1970s that industry stagnated and almost ceased.

Iron-smelting ceased at Blists Hill in 1912, the Coalbrookdale Company closed its Severn Foundry in 1917 and Coalport China Works closed in 1926. The remaining brickworks closed at the outset of the Second World War.[51] Along with this industrial depression went declining population, and a visibly deteriorating landscape. Slag heaps and pit waste littered the slopes of the Gorge, and Jackfield was almost buried under mounds of faulty red tiles. Even the wooded hillsides were scarred with open clay working and bald heaps of Pennystone waste. Yet industrial activity persisted.

The buildings of the old works were taken over by a series of small industrial enterprises.[52] They were often undercapitalised, and old buildings could be used as cheap premises. In Ironbridge, the old malthouses and warehouses of the riverside were reused by an assortment of tradesmen, including wire sieve-makers, the makers of iron patten rings for workmen's clogs, wiredrawers and ironmongers. Such industries served the needs of a local working population and were the forerunners of the small garages which appeared along the Wharfage and Dale End early in the twentieth century, and remain there today.

Such small workshop enterprises servicing local or at best regional needs are typical of most built-up areas. Capital investment was low, the element of skill high, and there was no need for elaborate transport arrangements. But industries supplying the needs of Ironbridge alone provided insufficient employment for those left out of work by the closure of old works. New industries were attracted to the area by a large pool of labour, by the availability of old buildings and by the planning policies of the two world wars. After heavy bombing, the London firm of Chillcotts Ltd moved into part of the old Coalport factory in 1941, where they made gaskets, washers and exhaust systems until the 1960s. In another part of the works Frank Hawker Ltd made fancy metalware powder compacts, lipsticks and trays. There was a new vogue for teddy-bears in the 1920s, and the Merrythought Company was set up in the old Severn Foundry.

Teddy-bears are still made in Ironbridge today, but many more of the industries which began life in the Gorge have since moved away. In the 1940s and 1950s, Ironbridge was effectively a 'nursery' for small industrial enterprises. Initially set up in old buildings, several – such as Clifford Williams making pyjamas – expanded to the extent that they were able to move to new purpose-built works elsewhere. In the old buildings of the Calcutts Ironworks, W.H. Smith began making machinery for the local clay industry in the 1870s and later aeroplane parts during the Second World War. They were taken over by Messrs Marshall Osborne & Co., and moved into the old Craven Dunnill factory. Since then, they have moved to new premises in Telford. What most differentiated these industries from the small workshops of Ironbridge or Broseley was their market. All were orientated either to a national or, in the case of Merrythought, an international market. The population simply could not sustain the localised industries which might be found in and around a large city.

Despite a falling population, a remote location and the closure of the larger works, industry had managed to survive. By pioneering new processes or products, and by keeping capital costs local, firms could establish themselves and, if they were successful, move away to more convenient sites. Today some of the older manufacturers remain, but in recent years few new firms have set up in the area. There has been a distinct slowing of industrialisation in the Gorge as a result of policies imposed by the New Town of Telford.

In January 1963, much of the Coalbrookdale Coalfield including the Gorge was designated as Dawley New Town by the Ministry of Housing and Local Government, and set aside for what was to be the largest new development in the British Isles. The aim was to attract overspill population and industry from Birmingham to a totally new environment. Sceptics argued that in fact the existence of old industrial buildings already created the preconditions for industrial expansion, and that the provision of new premises might have the opposite effect of deterring industry, through high costs.[53] In the early years, it seemed that these fears were justified; few industries came to the area, and at one point the rate of creation of jobs was falling. In 1968 the whole concept was reconsidered, a greater area taken in, and the town renamed Telford. With the completion of the M54 and the lifting of restrictive policies which brought in industry from a wider area, the industrial fortunes of Telford have revived. Ironically, by deliberately trying to encourage industries to come into the area, the planners had halted the process of industrial regeneration which was already well established in and around Ironbridge.[54]

As part of the planned new town, the Gorge was set aside as an area of amenity value for recreational purposes. The official image of the landscape had changed from one of squalor and dereliction to one of 'special quality'.[55] New industry was to be discouraged, and the emphasis moved from development to preservation. Industrialisation had, for the first time, been deliberately fossilised. At first the Gorge was seen as a local beauty spot, for picnicking and listening to the sound of running water[56] – derelict buildings were demolished, and a programme of tree planting undertaken.

At the same time another view of the Gorge was emerging. There was a growing awareness of the importance of the historic remains and in particular the Old Furnace at Coalbrookdale and the Iron Bridge, and a group was pressing for the establishment of an industrial museum.[57] It was recognised that there were many more sites and buildings of significance than could be housed in a single museum, and the remit widened to include a much broader concept of industrial archaeology. The idea of preserving an industrial landscape as a whole, rather than isolating relics in a museum setting was a new concept and not something which gained immediate acceptance. Nevertheless, by 1980 most of the Gorge was designated as a Conservation Area. Further demolition was regulated, and grants were provided for the rejuvenation of old properties. Industrial remains and old buildings had triumphed over trees as the official view of the area was transformed from one of local amenity value to one of international status as a World Heritage Site.

3

MINERAL RESOURCES IN THE LANDSCAPE

The landscape is a document. Imprinted on it are the effects of the natural processes of many millions of years, and of the more recent human activities of the last few thousand years. Surrounding us is a palimpsest of the past, its various changes superimposed one upon the other to create the pattern which we see today. This pattern is a record which documents these events, a record which is in itself a source of historical evidence for the past, in the same way that a written document charts the past. The landscape is a source of physical history in its own right, but, as with a document, there are many different ways of reading it. Like a document, the landscape has to be viewed critically and with great care.

Two aspects of the landscape which can be related to industrial development are the natural landscape and the man-made one. Inevitably the two become interrelated through time, but it is more convenient to begin here with the natural landscape. This chapter sets out the way one of the key features of the natural landscape – the underlying minerals – was exploited by industrialists, and the way in which those minerals, in turn, affected the pattern of industrialisation. The intention is not to put forward a crude model of economic determinism, but to explore the complex relationship between minerals and the wider economic pattern.

The geological landscape

The landscape of the Ironbridge Gorge is spectacular. The River Severn, meandering past the rolling countryside around Buildwas, abruptly enters a steep-sided, wooded valley. Passing through this, its course drops away over rapids, its banks alternately built up for wharfs, or cutting away massive heaps of industrial waste. Near Apley the surrounding agricultural lands once more slope gently down to the river bank. Exposed on either side of the Gorge as far as Coalport are valuable minerals – coal, clay, ironstone and limestone (fig. 3.1). Fast-flowing streams drop down into the river from the surrounding plateau, a potential source of water-power for early industry. Their progress has created a series of side valleys – Coalbrookdale, the Washbrook Valley, Benthall Valley and the Calcutts Valley. It was these minerals, the power supply, and the river itself as a means of transport that made the area such an attractive industrial location.

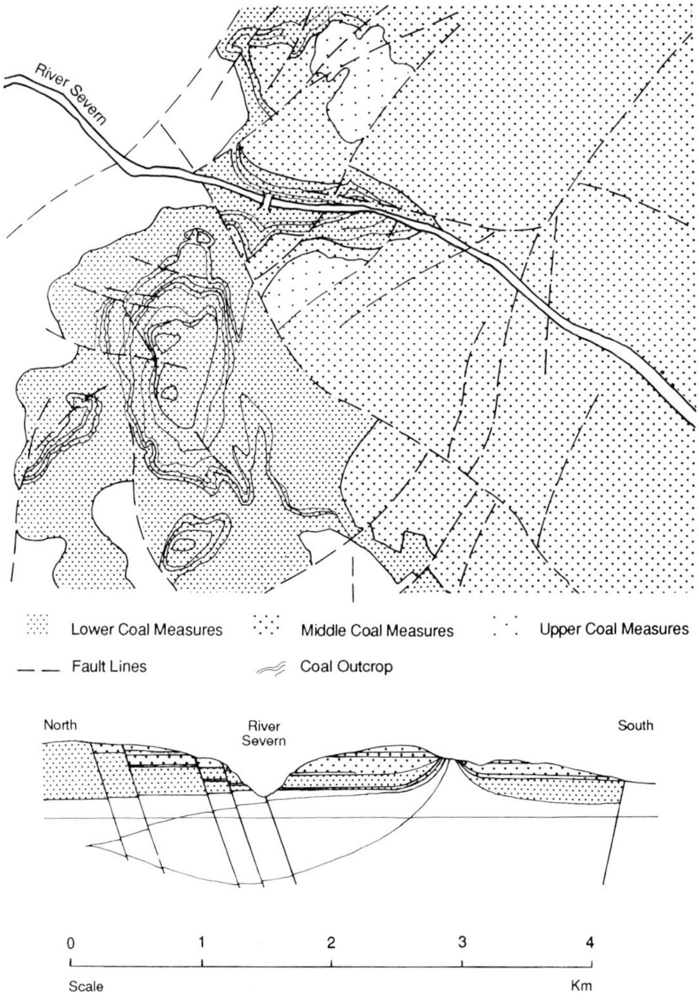

3.1 Mineral resources in the Ironbridge Gorge (Shelley White)

To industrialists, the minerals of the Coalbrookdale Coalfield were among the most important natural resources available to them. Coal, clay, ironstone and limestone were used by industry for over 500 years. But the pattern of mineral exploitation was complex – for example, it was not just a matter of digging out all the available coal until it was exhausted. There were many different types of coal, which served different purposes. As industry developed, so did the demand for coal, while at the same time new mining technology made new seams and reserves available to industry.

For other minerals, too, this constantly changing pattern of availability, technology, demand and understanding can be traced in the landscape. The exploitation of minerals is a classic case study of the way the human landscape interacts with the underlying geological one.

It is most convenient to discuss this landscape in terms of a geological resource: in other words, where different seams occur, and how they occur – whether deep beneath the surface or as outcrops. It is possible to interpret quarries and pit heaps, to predict where other mining areas might be found and to understand differences in mining techniques. At its most basic level, this information presents us with a general history of mining, but if we take a wider view of the landscape the underlying geology provides a single framework for the development of the area as a whole. Mining provides a context for the broad span of industrial development from the early opening up of the Gorge to the eighteenth-century iron industries and the specialist clay industries of the nineteenth and early twentieth centuries. A geological framework is a useful vehicle for understanding the field evidence for mining and industrialisation.[1]

Mining evidence as archaeology

As a class of archaeological evidence, mining has to be dealt with carefully, not least because of the inherent dangers of underground workings. Open-surface quarries may be easy to see but, by their very nature, they may obscure earlier workings. It is usually assumed that the visible surface pattern only reflects the most recent operations, with the implication that nothing earlier can have survived. Nevertheless, traces of earlier workings, or hints of their existence, may remain high up on the sides of quarries, or in the evidence for associated activities such as access roads or old spoil heaps. It is vital that even recent mining landscapes be treated in a careful archaeological fashion, and not dismissed outright on the grounds that earlier working has been lost.

Underground remains are much more difficult than surface workings for the non-specialist to explore. Tunnels and shafts are rarely accessible without specialist equipment and the means of monitoring gas levels. The pumping and winding equipment on which their operation depended has generally been lost, and large areas may be flooded or collapsed. Even if mines are not to be entered, it may be difficult to locate mine entrances, which decay very rapidly once they go out of use. Of the many tunnels known in the Gorge, only one or two are still visible today.

Other field sources of information for mining are indirect. Waste in fields or around old shafts may be visible as heaps or as stains in a ploughed field, perhaps visible in aerial photographs. Unfortunately for archaeologists, many local mining techniques such as long-wall mining used old spoil to prop up the roof and so did not produce much surface spoil. At other mines, spoil was dumped into the river and washed away. A second source of surface evidence is patterns of subsidence, which can reveal old or inadequately capped shafts, the presence of workings on an outcrop, or underground workings close to the surface. The amount of subsidence relates to the working methods, the roof material and the seam thickness. Typical patterns might be shaft collapses, crown-hole incidents associated with outcrop or shallow workings, and general trough subsidence.

Where they cannot be seen, mineral workings can be identified on deposited plans, but legislation requiring public deposition was introduced in only the 1870s, so the survival of mine plans from before that date is a random matter. The earliest Ordnance Survey maps which routinely show shafts or recently abandoned shafts date to 1883, which is very late in the history of mining in the Gorge. More useful is the recent 1:10560 Geological Survey sheet SJ60SE, which collates historical information about shafts, and includes the results of trial-hole borings. Information about old workings can be found in these. Geological cores may reveal voids, or the presence of 'gob' or waste from old workings, and can also be a useful source of information.

Once identified, mining remains are at risk of treatment in the interest of public safety by such techniques as concrete capping or, more drastically, total infilling by grouting. These can render mines permanently inaccessible.

The Silurian landscape

The most dramatic and visible evidence for mining in the Gorge is the remains of the limestone industry. Towering above the Iron Bridge is Lincoln Hill, pale yellow and scarred by open quarrying for limestone.

The limestones of the Silurian Formation occur in a ridge across the western end of the Gorge, and were used for iron-smelting, building stone, lime mortar and cement. Four hundred million years ago, equatorial Shropshire lay under the large, shallow Iapetus Ocean.[2] The climate was warmer than today, and extensive coral reefs formed on the floor of the ocean. The gradual transgression of this sea formed the Silurian limestones, known as the Wenlock Shales. The Silurian limestones underlie the later Carboniferous deposits, but through faulting have been pushed up, and are exposed on the surface as a high ridge crossing the valley from north to south. On the north is Lincoln Hill, pushed up by the Lightmoor Fault and denuded of its covering coal measures. On the south bank, Benthall Edge is a continuation of this ridge, turning south-west to become Wenlock Edge eventually.

The limestone itself is of variable quality. The uppermost band of Wenlock Limestone is good-quality stone, and contains many of the prized 'ballstones' or 'crog balls'[3] – residual lumps of pure coral which create a very high-quality stone. The surrounding and lower measures are of lesser quality and rest on the very poor-quality lower Benthall and Tickwood Beds. The uppermost levels were used as fluxing stone in iron-making, to draw out impurities during the smelting process. This stone could also be used for building, but the local material could not be cut smooth, and is more commonly used for rubble fill, or rough cottage walls in and around Much Wenlock. The lower, poorer quality stone was burnt for building or agricultural lime.

The limestones of Shropshire played an influential role in the history of geology in Britain. Shropshire was a focus of very early geological interest, partly because of the economic importance of its minerals, and partly because of its relatively simple geological sequence. It was through studying the fossils in the limestone strata of Wenlock Edge that Murchison was able to develop his great Silurian System, one of the first geological sequences to be fully defined. His book, *The Silurian System*, first published in 1835, generated a great deal of controversy over the dating of geological measures and was eventually to become instrumental in the revision of ideas about the age of the world. Hamstrung by an antediluvian explanatory framework, the published academic understanding of geology developed very much more slowly than the practical one.

There is evidence for the exploitation of the Silurian limestones on Benthall Edge, dating back to the thirteenth century. Benthall Edge is a 100-metre-high scarp of limestone on the south bank of the river. The cliff is capped with high-quality limestone, which is needed for fluxing; and lower deposits of poorer-quality stone could be burnt for lime and building mortar. Along the top of the ridge is a long string of limestone quarries, with associated spoil heaps, roads, railways and limekilns (fig. 3.2). The Edge is a classic example of an industry which has left very extensive remains in the landscape, but for which very little documentation has ever been found.

3.2 View of the Gorge from Patten's Rock Quarry, 1856. A truck loaded with limestone and the gear for an incline may be seen in the foreground

In order to understand this industry, a detailed survey of all the remains was undertaken.[4] Over 100 quarries, roads and limekilns were identified and mapped as part of an intensive archaeological survey. Through the use of an archaeological matrix which recorded the relationships between quarries, it was possible to analyse the landscape evidence for the history of limestone mining here (fig. 3.3).

To the west, medieval quarrying is linked with the very early road to the nearby Cistercian abbey of Buildwas. The quarries feed out onto a road which zigzags down the steep escarpment to the abbey. These workings have been cut by later workings, and there is no evidence that they continued in use.

The second phase of working can be linked to the iron industry. The quarries that run along the very top of the scarp provided the fluxing stone needed in the smelting of iron. These quarries were working the high-quality ballstones and uppermost reef limestone, which was acknowledged to be 'of the best quality for smelting iron'.[5] The stone was removed via a group of roads and inclines which converge at a point on the river bank where a wharf was located, known as the Stone Port. Few of these quarries have limekilns, suggesting that relatively little of the stone was burnt *in situ*, although there was a pair of kilns by the wharf. This phase of exploitation represented the transport of fluxing stone by river to supply the early eighteenth-century iron industry at Coalbrookdale. Documentary evidence refers to barge owners shipping stone to Ludcroft Wharf at the foot of Coalbrookdale shortly after 1717.[6]

Another series of long, open quarries is linked with a railway, leading in the opposite direction, away from the river and towards the fields of Benthall and Broseley. Twenty kilns are known from map and field evidence all relating to this railway. Most of this stone was used for agricultural lime. One set of kilns was new in 1800 and the railway was still in use for lime-carrying in 1844.[7] The railway was connected with Benthall Rail, a wooden, horse-drawn railway in existence by 1686 which runs down to the River Severn. Finally, at the eastern end of the scarp two inclines lead down to ten limekilns on the river bank. Initially the lime was taken by river but when the quarries were reopened in *c.*1926[8] another incline, crushing machinery and a railway siding were built. The pattern of quarrying reflects a move from exploiting high-quality stone for fluxing to lower-quality stone for lime and agriculture. Yet the better stone can still be seen in the quarry face, and has by no means been exhausted. The exhaustion of limestone was perhaps not the only explanation for the demise of working.

There were also extensive workings on Lincoln Hill, where the great open gash of a former quarry can still be seen above the Iron Bridge.[9] In fact, only a small proportion of the originally extensive works can be viewed today.

A more superficial survey of the works than was possible at Benthall suggests a similar pattern – until *c.*1760, the working of high-quality limestone following the outcrop of the stone. Only suggestions of this early phase survive, including the quarrying of the area around the Swan public house.

Once the surface stone was exhausted, working continued underground. On the south side of the river there is no evidence for underground working, but here the measures slope downwards beneath the town of Ironbridge. Below the hill were hollowed-out caverns, their roofs supported by pillars of remaining stone in a technique known as 'pillar and stall' working. The technique leaves large voids, and often results in surface collapses, which is indeed what has happened in the vicinity of

Lincoln Hill. Because of the danger posed to the road, the workings have recently been filled in by the County Council, and are no longer accessible.[10]

The final stage of working on Lincoln Hill took place during the nineteenth century, when the poorer-quality remaining stone was worked for building and agricultural lime, a phase which coincides with a building boom in Ironbridge itself. Of the 32 kilns shown in 1842, only three survive – two are preserved on the Ironbridge Wharfage, and above them lie the collapsing remains of a third kiln. Once again, what survives today represents only a small part of the industry at its greatest extent.

This latter kiln was almost certainly one of the kilns painted by Turner. The Lincoln Hill workings were very much part of the drama of the industrial landscape; they were also painted by Munn, and mentioned by many visitors, including Perry, who described the hill as a 'frightful precipice' and Katherine Plymley, who admired the 'grand, rude arches' of the caverns.[11]

Benthall Edge and Lincoln Hill were among the closest sources of limestone for local furnaces, but most of the ironmasters were forced to obtain fluxing stone from more distant sources. The Coalbrookdale Company bought limestone from Much Wenlock and Gleeden Hill, 4 km beyond Benthall Edge. The partners in the railway to the Stone Port at Buildwas (who operated the furnaces at Bedlam, Calcutts and Wren's Nest) obtained stone from Wyke and Tickhill, and not from the nearby Edge.[12] By 1835 the better stone had been worked out, and the furnaces were using stone carried from Llanymynech on the Welsh border along the Shropshire Union Canal.[13] It seems that the Harries family, who owned the mineral rights to Benthall Edge, were content to allow the quarrying of poorer-quality shales for lime burning, but did not provide limestone for competing furnace owners, however profitable this might have been.

The significance of limestone is often underplayed, partly because the industry has generated so little documentation[14] and partly because remains have been lost or obscured. As with other mining operations, working did not cease once the better-quality stone had been exhausted and the boom period of the Industrial Revolution had passed. The growing demand for agricultural products generated by industrialisation and the rapid construction of housing in Ironbridge,[15] combined with the practical difficulties of keeping burnt lime dry, created an extensive market for lime which was filled locally throughout the nineteenth century.[16]

The Carboniferous landscape

The coals, clays and ironstones of the Carboniferous measures were the chief source of the Gorge's economic advantage in the seventeenth and eighteenth centuries. But their exploitation was neither straightforward nor simple.

During the Carboniferous period Shropshire was covered with eroding river channels, interspersed with large areas of swamp and lake. The layers of peat formed from swamp vegetation became the coal seams, while tectonic movements led to occasional marine incursions, creating the marine bands in which ironstones are found. The

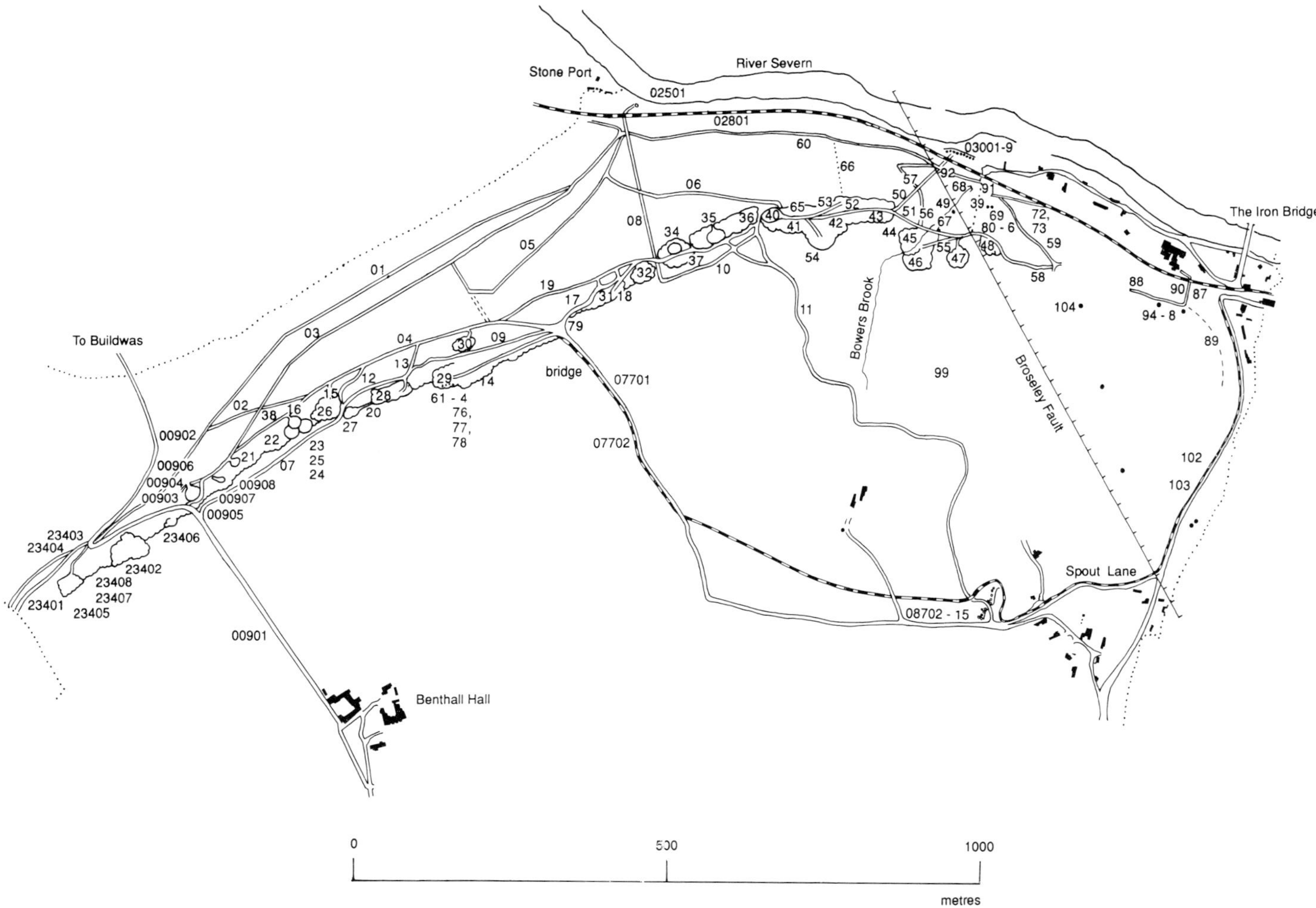

(*a*)

3.3 The archaeology of the Benthall Edge limestone industry: (a) plan of quarries and (b) stratigraphic matrix

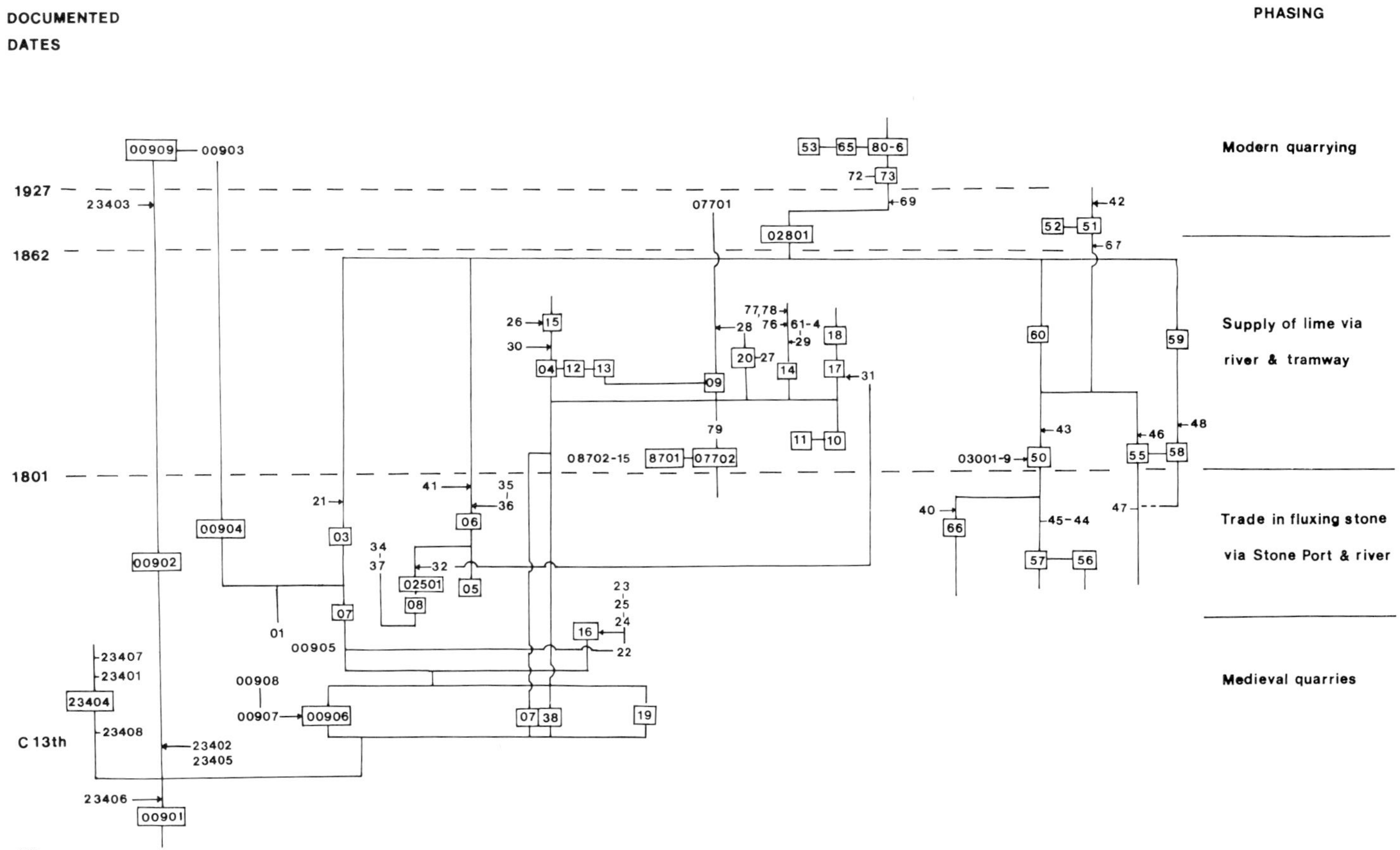
DOCUMENTED
DATES
PHASING
1927
1862
1801
C 13th
Modern quarrying
Supply of lime via
river & tramway
Trade in fluxing stone
via Stone Port & river
Medieval quarries

(*b*)

Shropshire Coalfield extends almost 20 km from Lilleshall in the north to Willey in the south and is conventionally divided into three: the Lower Coal Measures, the Middle Coal Measures and the Upper Coal Measures. The Ironbridge Gorge itself was created very much later during the last glaciation. When the ice sheet of the last glaciation melted, a large lake formed between the watershed of the Wenlock and Pennine ridges, and the face of the ice sheet. As the water-level rose, the lake overflowed at the Ironbridge col, cutting through the coal measures and limestones. The resulting Gorge still retains the characteristics of a recently formed valley – the sides are steep, and fast-flowing streams have created side valleys. The cutting provided an easily accessible section through the principal Carboniferous minerals – iron, coal and clay.[17]

Coal

The importance of coal in the process of industrialisation has been shown in chapter 1 – a seventeenth-century industry with medieval origins can be seen to underpin the earliest economic development of the Gorge. In the eighteenth century coal-mining and the discovery of new applications for coal stimulated industrial diversification, and by the 1750s coal was integrated with local ironworking as the industry moved from export to local consumption, the miners themselves providing an industrialised population which adapted well to other industries. Coal-mining continued throughout the nineteenth century and into this century, largely as an adjunct to clayworking but still with an economic role to play.

Landscape features associated with coal can be clearly seen around Ironbridge. Groups of houses such as 'The Mines' and Broseley Wood began as coal-mining settlements, remote from the traditional agricultural town centres. The pattern of industrial housing is a product of the needs of miners, and this in turn generated a local brick industry. The first transport links down into the valley were stimulated by the need to get coal to the Severn, and coal supplies were vital to ironworks such as Bedlam. The remains of the many industries which relied upon coal supplies have shaped the development of the south bank of the river. The need to maintain supplies of coal generated external transport links such as canals and railways, after local supplies had been exhausted. And the legacy of coal-mining today remains in the areas of poor soils, and in the disturbed and slipping ground which is not economic to develop.

But the basis of all this activity was the coal itself – the seams of the Middle and Lower Coal Measures. In order to understand the development of the coal industry, it is necessary first to be aware of the properties of these seams and the way in which they occur on the ground.

In sequence, the lowest measures are the Lancashire Ladies and the Crawstone Coals, which were only ever mined from shafts. The lowest seam to outcrop on the surface was the Little Flint Coal. Above this were the Best, Randle and Clod seams – some of the most highly prized coals which outcrop on the river bank between the Iron Bridge and the Free Bridge, and on the surface in Benthall parish. Field survey in Benthall parish has shown that there is a consistent link between outcrops of the coal and surface evidence for early workings. Above this was the Ganey Coal – a much-worked seam probably in association with the Ganey Clays – and the poorer-quality Clunch or Viger Coal. The uppermost measure in the Lower Coal Measures was the New Mine Coal – a highly sulphurous coal which was rarely worked until the nineteenth century. The Big Flint Coals of the overlying Middle Coal Measures were also worked.[18]

Early industrialists were highly selective about coal seams. Long before the properties of coal were documented, local miners distinguished the different strata. Low sulphur content was the main quality sought in the seventeenth century and all but the Main Sulphur and the New Mine seams were suitable. It is to these low-sulphur coals that Darby's success in smelting iron with coke has commonly been attributed.[19] By the mid-eighteenth century a more critical distinction amongst local coals emerged – it was universally recognised that the only coal suitable for smelting was Clod Coal.[20] With improved iron-making technology higher sulphur levels could be tolerated. By the mid-nineteenth century, the use of coal for the railways, pottery, brick- and tile-making, and in gas production, was on such a scale that smelting was only a very small proportion of its use. These new uses meant that seams once thought to be exhausted were reopened[21] – a process that continued to the present day with open-cast mining of worked-out coal seams and coal wastes of very low grade, for use in electricity power stations. Mining evidence has to be understood in terms of the demands of industry.

The physical effect of the Gorge itself was to cut through these coal measures, exposing them to the surface on either side of the river. In other areas such as Benthall and Broseley town, faulting and erosion brought coal close to the surface. Thus coal was accessible in the area before the development of the complex mining technology of the past two hundred years. These two factors – quality and access – were instrumental in the development of coal-mining here earlier than in comparison to other mining areas in Britain, and continued to govern the pattern of working.

The third factor in understanding the pattern of exploitation was the level of resources available to miners. The choice of mining technique is a factor of the scale of operation and the capital available. Broadly, mining techniques developed from medieval open quarries of coal outcrops, through bell pits and hillside adits, to deeper shafts and long-wall mining. However, in reality methods were thoroughly *ad hoc* and depended more on individual circumstances and the investment available than on the inevitable march of technological progress.

The classic landscape pattern of shallow bell pits – depressions surrounded by circular mounds of spoil – survives on the south bank at Workhouse Coppice in Benthall, Deerleap Woods in Broseley and in the woods at Caughley. Shallow pits occur where the coal is relatively close to the surface, and those that are preserved clearly follow the seam outcrop. At Caughley, one band of bell pits reflects the outcrop of the Best, Randle, Clod and Ganey seams, and a second, further down the slope, the Little Flint Coal.[22] Dating bell pits is almost impossible, and it can only be assumed that the majority of Benthall pits are mid-seventeenth-century in date, when coal production was at its peak, and that those at Caughley are pre-1750, when they were cut by later clay workings.

Part of the problem in dating bell pits – and indeed other types of mining – is that there was not a simple temporal transition from the use of bell pits to deeper mining. Rather, the mining technique seems to have been dependent upon available resources. Landowners such as James Clifford in 1575 and John Weld in the 1630s were able to sink deep pits which presented a problem of waste disposal and required pumping, at the same time as individual miners were working single bell pits. Even the open working of coal on the surface – for which there is evidence at Linley, at Bagleys Rough and other areas – is undatable. It can be assumed to be the earliest technique used in the medieval mines, and yet coal was worked from surface outcrops during the strike of 1912 and at Woodside much more recently.

The way in which coal seams outcropped on the sides of the valley dictated that one of the most common techniques was the adit or insett – a tunnel dug directly into the seam from the hillside, with smaller galleries leading off it. The technique was common by about 1608, when James Clifford dug 'fower insetts' at Ladywood (figs 3.4 and 3.5),[23] and was still in use for getting fireclay and coal from above the white brickworks at Benthall in the 1950s. What was notable was the length of these insetts in the seventeenth century – 1000 yards and longer.[24] Unfortunately, adits have a notoriously poor survival rate as field evidence. The Benthall adits just above the Iron Bridge ceased working in 1953, and were visible in the 1960s, but today little more than a depression in the hillside survives. Clifford's adits were mapped in *c.*1621[25] but despite archaeological excavation there was nothing visible in the disturbed river bank where they ought to have been found.[26]

In the second half of the eighteenth century there was a clear transition to mining on a company basis, as part of the integration of ironworking operations. Companies purchased mining operations, such as the limestone workings at Lincoln Hill, which were bought by the the Madeley Wood Company, or the investors in the Willey furnaces who bought coal-mining operations. These companies could command large amounts of capital, and so there was a corresponding change in the scale and techniques of coal-mining. The Broseley Hall Estate Map of 1728/65 shows that there were several pits in Jackfield mining huge areas of coal by the long wall method (fig. 3.6).[27] A long, continuous working face is kept open, and the space behind filled with gob to support the roof. Modern workings frequently encounter old gob in coal seams, and the practice may explain why so few pit heaps from early workings can be identified.

By the mid-eighteenth century, both adits and shafts were large enough to require more elaborate technical investments, including ventilation and mechanical means for forcing out water. Earlier mines had had air shafts, and 'soughs' for drainage, but little evidence of these survives. The Ladywood Sough comes out alongside the old railway track below the hairpin bend on Broseley Bank and is still flowing copiously, a very ochrous water. However, at Benthall a small brick structure survives which may well have been the base of a chimney which created a through draught to ventilate an adit. One of the rare surviving pieces of field evidence for pumping is at

3.4 Extract from Samuel Parsons' Map *c.*1621 showing insetts or mines on the riverbank (SRO 1224)

the Lloyds Coppice on the north bank of the river. There two shafts, pump rods and brick foundations survive from a pumping and winding engine. This could well be the earliest in the Shropshire Coalfield, erected by the lessees of the Madeley mineral rights in 1719.[28]

Company investment in mining is also visible in the remains of the Heslop engine base, excavated at Blists Hill.[29] There the Madeley Wood Company erected a Heslop-style engine in the 1790s to pump water from a shaft. Heslop engines were a variant on the Boulton & Watt engine, developed by a local man and widely used in the area. Two characteristic 'haystack' boiler bases remain, but the rest of the engine base has been altered by the insertion of a later engine.[30]

The Tar Tunnel at Coalport is the only safe and easily accessible mine in the area.

3.5 Lateral mineshafts beside the Severn, Dibden 1801 (IGMT SSMT18)

The Madeley Wood Company proposed a novel solution to removing coal from the base of their shafts – a tunnel driven in from the riverside to the bottom of the shafts.[31] Soon after it was begun, the tunnellers struck a source of mineral tar, and the operation was diverted to tar extraction. It was only after the tar was exhausted that the tunnel was taken through to the coal shafts, where it acted as a drain. The tunnel remains open to the public today as part of the museum complex, and can be explored for several hundred yards.[32]

If the eighteenth century was a period of innovation and investment in local mining, the later nineteenth century stands in contrast as one of 'make do and mend'. The pattern of company mining by large ironworking concerns had given way to one where coal-mining was mainly an adjunct to brick and tile manufacture. Easily accessible supplies of the famous Clod Coal had been exhausted, ironmasters had developed new sources of coal and ironstone in the northern part of the field and most of the market for Broseley coal was local. By 1840, the majority of shafts in Broseley were worked by brick- or tile-makers, who were producing coal for their own use, and coal-mining had ceased altogether in Benthall.[33] The high number of accidents in local mines reflects the rather casual nature of mining technology.[34] Although pits were generally deeper and more extensive than they had been, old

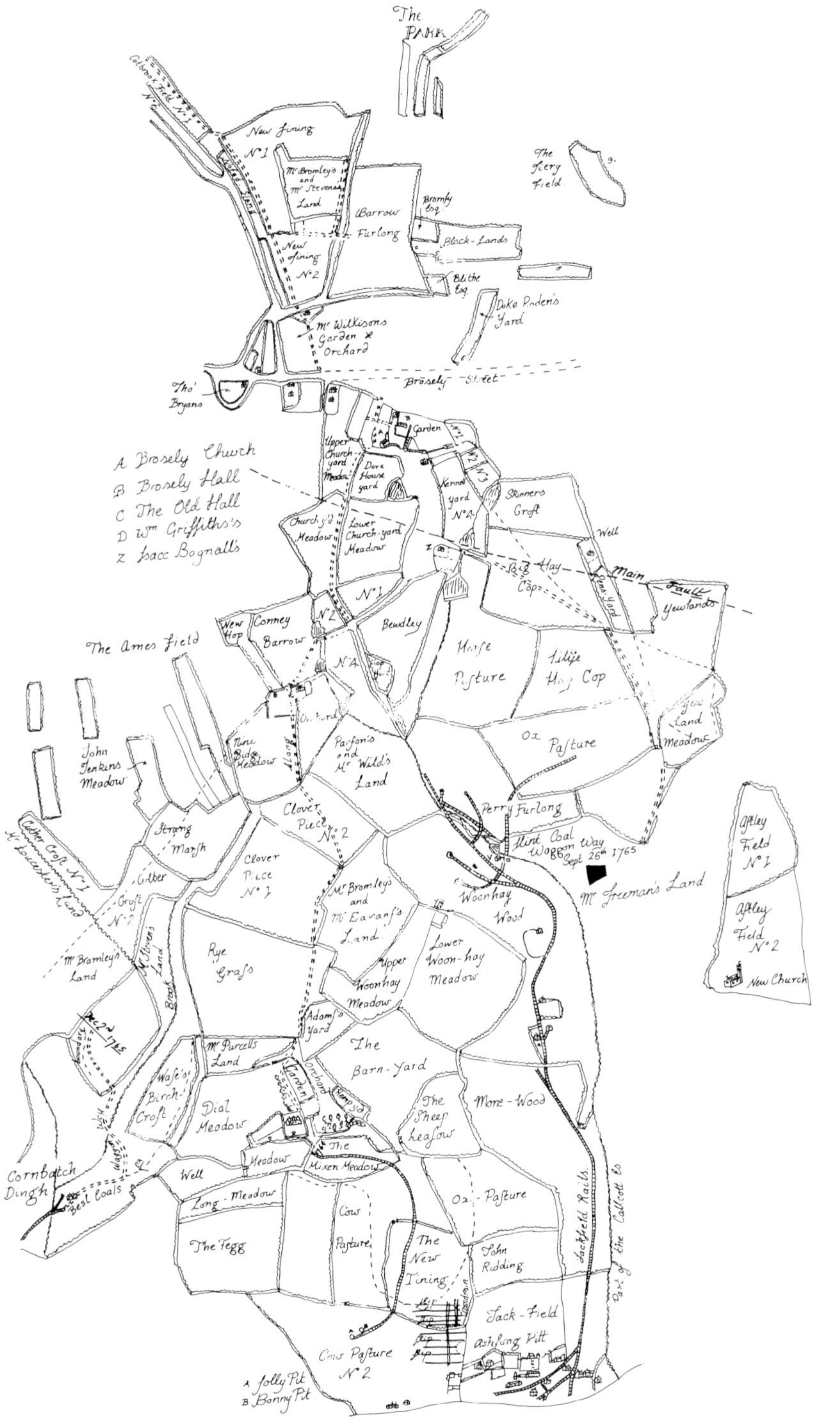

3.6 The Broseley Hall Estate Map showing early evidence for railways, pits and also limits of underground working. Based on SBL MSS, *c.*1728/65 (Nicola Smith)

engines, such as the one at Deep Pit in Broseley, remained in operation, often working shafts up to 150 years old, and hand winding was common.[35]

Although the main seams were said to be worked out, sporadic mining did take place earlier this century. Clod Coal was certainly being worked at the Viger mine in Benthall in 1953 where the remains of adits and an inclined plane over the railway line are just visible,[36] and several old pits were reopened around Benthall hall during the Second World War. More recently there has been renewed mining activity through open-cast working. The Ironbridge power station has generated a market for local coal which need not be of particularly high quality, and this was supplemented by the supply of clay to brickworks in Telford. At Caughley and the Deerleap, where mining was abandoned earlier this century, new open-cast techniques on a vast scale have made the seams once again economical.

Thus, while the quality of coal and the ease of access to it lay behind the pattern of early coal-mining, later mining was a complex and changing process, dependent not so much on adopting the latest mining technology, but on the amount of investment industrialists made available. Coal-mining moved from being a dominant industry in its own right to becoming simply an adjunct to other industries, and in doing so lost the ability to attract investment.

Ironstone

Of the many ironstone bands found over the north of the coalfield, only two occur significantly in this area – the Pennystone and the Crawstone.[37] The most commonly exploited was the Pennystone, consisting of small, smooth nodules of iron carbonate in a band of mudstone, sometimes capped with irregular masses of ironstone. It was one of the most valuable of the ironstone seams.[38] It outcrops on either side of the Gorge around the 100 m contour and lies under the Big Flint Sandstone, which often provides a good stable roof to mines. Pennystone was brought to the surface in its clay matrix, which was left to weather and the lumps of ironstone picked out, often by women (fig. 3.7). The nodules were then calcined, by being roasted in heaps, and later in specially designed ovens.[39] This reduced the total bulk of the ore, removed water and impurities, and was normally done at the ironworks prior to charging the furnace, rather than at the mine itself.

Crawstone – a sandstone impregnated with ironstone – occurs near the bottom of the Lower Coal Measures and was mainly obtained by deep mining, but it outcrops in Madeley Wood, and possibly elsewhere near the riverside in Ironbridge and Benthall. It is the only ironstone measure found south of Broseley, where it is said to be of better quality, and there was extensive evidence for mining it in the late nineteenth century.[40]

From an early date, ironmasters clearly appreciated the distinction between these two types of ironstone. Thus the 'Baull' stone and 'Flatt' stone bought by Abraham Darby I in 1708 were strictly differentiated in quantity and price. They indicate a degree of deliberate mixing of stone even at an early date.[41] The reason that such care was taken to select the right ore was because of the problem of phosphorus. Most

3.7 'Adams Engine and Ironstone Pit', Madeley Wood 1847. Women may be seen to the right of the image picking ironstone nodules out of the clay matrix (IGMT1992.9785)

impurities in iron-ore are lost during smelting; phosphorus remains, producing an iron which is fluid in casting. The phosphorus levels in local ore may explain Abraham Darby I's concern at selecting the right ore, which would allow him to produce thin castings.[42]

Unfortunately, pig-iron high in phosphorus is 'cold-short' or brittle, and unsuitable for making wrought-iron, thus limiting the use of early coke-furnace iron to relatively crude castings. The recognition and selection of low-phosphorus ironstones (specifically the Ballstone, Blackstone and Pennystone) by Abraham Darby II in the mid-eighteenth century made it possible to produce pig-iron suitable for making wrought-iron in a standard charcoal finery.[43]

Ironstone mining techniques were similar to those of coal, with a broad, but by no means absolute, succession of bell pits, replaced by shafts and adits. Coal and iron were often mined together, although on the north side of the river, where thicker ironstone seams were found, ironstone continued to be mined as a separate operation. A Crawstone mine in Ironbridge, operating as late as 1842, worked a circular long wall face, accessed through three tunnels. About 0.6 m of massive ironstone was worked, as well as the thin seam of coal on the floor. Evidence for the techniques used in working the Pennystone comes from mines to the north, where thicker seams were worked in different levels.[44]

From the surface, the key to identifying former Pennystone mines is the large, often bare spoil heaps and the ochrous water running off the mine outlets. Pennystone mines produced at least 10 tons of spoil for every ton of iron produced, and in order to remove the iron nodules more easily the spoil was left to weather on the surface.

The clay was apparently 'injurious to vegetation' due to the salinity of the deposits, and so the mounds remain for years without grass. It broke down to a clay which was occasionally used in brickmaking.[45] The most obvious heap is at the north side of the Free Bridge, where a vast flat-topped, bare mound is approached by a steep track cut into its side. Two little stone cottages have been built on the top – almost certainly squatting on otherwise abandoned and useless land.[46]

Several Pennystone mines have been located in Ironbridge, although only one is open, but because of bad air it has not been fully explored.[47] There were also Pennystone mines on the south bank of the river in Ladywood, but only the spoil heaps remain visible on the surface today.

Crawstone mines are less easy to identify on the surface, as they do not produce the same quantity of waste. At least one Crawstone mine remains in Ironbridge. It consists of a tunnel with three main passages branching off it, and was worked using the long wall system. It seems to have operated between *c.*1823 and 1840, and the stone was sent by inclined plane down to the riverside below.[48]

Ironstone mining on the south bank remained important well after the smelting of iron had ceased in the Gorge. By 1840, James Foster had purchased land and transport networks in Jackfield, and was using them to transport iron-ore to the riverside for export to the Black Country. There were apparently calcining kilns on the river bank, although no evidence for these is visible amongst the tile waste tips of Jackfield.

At Blists Hill, however, there are the remains of two features relating to the calcining of ironstone. To the south of the furnaces is a level area, cut as a terrace into the hillside, with a brick wall retaining the canal embankment. Large quantities of brittle, russet-coloured ironstone finings can be found scattered on the ground, suggesting that this was the site of open calcining heaps, where ironstone was roasted to drive off impurities. At the rear of the furnaces themselves are the remains of a purpose-built calcining kiln in the form of the circular kiln mouth.

The mining of ironstone has left few visible remains. Yet the presence of ironstone must have been one factor in the siting of the early iron industry, and the physical conjunction of coal and ironstone – often in the same shaft – must have been instrumental in the development of coke-smelting.

Sandstone

The bands of sandstone found between coal measures also proved to be a useful local resource. Coal-measure sandstone was one of the earliest local building materials, and survives in early phases of houses in Benthall, Broseley and Ironbridge (chapter 6). It may have been used in early buildings at the Coalbrookdale works (chapter 5) but, apart from the Great Warehouse at Coalport, was rare in industrial buildings except as a foundation material as in the engineering buildings which remain today at Coalbrookdale. Sandstone was often used where there was contact with water (e.g. the riverside walls at Coalport or in Ironbridge)[49] and also in retaining walls against earth banks. Sandstone retaining walls are a key feature of the steep lanes of Ironbridge and along the line of the former Coalbrookdale railway. Sandstone was also employed for furnace hearths prior to the introduction of brick, and to provide loose sand for furnace castings.

The measures which produced sandstone were the Big and Little Flint rocks. Evidence for sandstone mining can be seen in local names such as 'The Quarry' in Broseley Wood, thought to be the medieval source of sandstone for Buildwas Abbey. There is field evidence for sandstone mining in the vicinity of Lodge Farm, Ironbridge. Stone and sand were transported from here down Church Hill to the Coalbrookdale Company works. Sandstone mining was also noted in exploratory boreholes near Bedlam Furnaces.[50]

The abutments for the Iron Bridge required two types of stone – rubble stone used to fill the abutments, and wrought ashlar facings.[51] Some of the stone came from the garden of Charles Hornblower in Ironbridge, which the Bridge Proprietors were obliged to make good in 1787. It is likely, therefore, that some of the terraces and areas of Ironbridge have been created by quarrying for sandstone.

There were other mining ventures which are more difficult to trace archaeologically. Salt brine exuded from the Flint Coal in Madeley parish,[52] and also was thought to have been extracted at the Salthouses in Broseley. Mineral tar was worked at Coalport and Coalbrookdale.

Clay

Clayworking began much earlier, and lasted much longer than any other industry in the Gorge. Bricks, tiles, pots and clay pipes were made in the area over three hundred years, and perhaps longer.[53] Even the established iron partnerships such as the Coalbrookdale Company and the Madeley Wood Company include brick and tile manufacture as part of their output. We tend to assume that iron and coal dominated the industrial prosperity of the area, yet in 1835 many of the inhabitants of Benthall were employed in the potteries, and brick- and tile-making remained an industry of regional and most likely national importance well after the demise of iron-smelting in the Gorge.

Clay is a bulky raw material, and bricks themselves represent relatively low added value in comparison to, for example, pottery, where the profit for the amount of clay is much higher. Transportation costs were therefore much more significant in the heavy clay industries; to minimise costs, brick- and tile-works were located as close as possible to clay sources. Thus, the industrial sites cluster at or around the main sources of clay.

Clay is found in a series of different seams between the bands of the coal measures, but each of these seams has very different properties. In order to understand the variety of clayworking which took place, and where it took place, it is important to understand the clay itself. The clay industries were dependent on three very different types of clays with different properties – the surface clays, the red clays (including the tile clays), and the fireclays.

Surface glacial clays were probably the first source of clay used to make bricks in the area, a pattern which has been identified in other parts of Britain.[54] The earliest dated brickwork here is seventeenth century.[55] In general, such bricks are red but occasionally burnt blue. They are thinner than later bricks, and slightly distorted. In section they are often coarse, with pebbles, air holes and other inclusions. The mixing of the clay is irregular, and bands of white and red can often be seen. Common house-bricks do not necessarily require specialist clay, and in the case of these early bricks the colour seems to be a product of the firing rather than of any individual clay source. Such bricks were probably fired at a low temperature in an open clamp, perhaps using wood as a fuel: over-fired bricks burnt blue, the remainder a soft red colour. These brickmaking operations were small, and often impermanent. Itinerant brickmakers such as Jos. Whitefoot on the Willey Estate would have set up their kilns at or near the building site to fill an order. Others set up kilns in populated areas, again to be close to builders rather than clay sources. One lease mentions the 'liberty to make brick on the premises for the use of the premises only'.[56] It is unlikely that such operations would have mined their own clay, or bothered to transport clay long distances.

Local surface clays would also have been used in the individual brick kilns to be found scattered amongst the farms throughout the period, often indicated on maps as 'Brick Kiln Field or Leasow'. Only at the Dunge Brickworks to the east of Broseley was surface clay exploited commercially. There, a fat grey clay was found, which made a characteristic yellow to grey brick with black spots, which can be recognised on some Broseley buildings today.[57]

The red clays were concentrated in the Coalport and Hadley formations of the Upper Coal Measures, which outcrop to the south and east of Jackfield and around Blists Hill. Their most important properties were a high iron content, and a degree of plasticity which enabled them to be moulded, making them very suitable for brick and tile manufacture. Within this category, there were two distinct types of clay – the clays used in making roof-tiles, and the softer red-burning clays.

When they were under threat from competing roof materials in the early twentieth century, the manufacturers of Broseley roof-tiles made much of their unique clay.[58] They argued that the quality of their tiles was totally dependent upon a clay found only in the Broseley area. The clay was found deep in the Hadley formation, and thus could not be open-worked. Because it was so hard, it had to be mined by blasting, and once on the surface required weathering to break it up. Such processes made it more expensive to get than open-worked surface clays. The tiles it produced were characteristically dark purple or brown, and very hard, since the clay was able to withstand very great heat in the kiln. The clay was sufficiently plastic to be moulded, but did not shrink or warp during firing, and produced an even, regular tile.

The deep mining of tile clay must have been established by 1776, when Arthur Young saw 'great numbers of blue tiles ... burnt here, and sent by the Severn to a distance'.[59] The Hollygrove and Ladywood works and probably several others were certainly founded in the latter part of the eighteenth century. By 1840 the Tithe map shows a series of large concerns already functioning on the riverside at Jackfield, characterised by a heavy investment in buildings and a scale of operation very unusual for the brick and tile industry. These works are nominally described in the apportionment as 'brick kilns' but they were almost certainly manufacturing roofing-tiles for export as well. Each of these works was located by the riverside, away from centres of population, and most had a path and loading area on the riverside itself.

By far the best archaeological evidence for the industry comes from the preponderance of purple/brown roof-tiles on the roofs of the Gorge and on buildings towards Bridgnorth and Shrewsbury. Very few buildings survive from these works. None of the shafts which provided the deep clay can be explored as they are either flooded or have been capped. Tiles are almost the universal roofing material, with the exception of a few slate-roofed buildings dating to the period after the construction of the railway line.

The red clays of the same formation could not take such intense heat, and burnt red in the kiln. They were ideal for floor-tiles, producing a fine-grained product which would gradually wear smooth. The products had to be dried very carefully and slowly, or they would be liable to crack during firing.[60] Clay floor-tiles are often found in eighteenth-century buildings, the earliest of them seem to be stamped 'I.B. Davis'.

As the clays were found in the same formation, firms producing roof-tiles often also produced red clay products. By the late nineteenth century, most of the Jackfield firms made a standard range of products including roof-tiles, floor-tiles and decorative finials, in red, blue, brown and white, suggesting that they had access to different sources of clay. So far no evidence has been found for the manufacture of earthenware pottery from red clays and, although the famous Jackfield ware has a red body, wasters of this ware have not yet been found. Red clays were, however, used at Benthall in the twentieth century for black-glazed electrical goods and lampstands, and seem also to have been used for art pottery.

The important issue is whether the clays were responsible for the demise of the Broseley roof-tile industry in the late nineteenth century. Various culprits have been blamed for the problems manufacturers faced from the 1880s onwards, including competition from slate, and 'bacteria' in the clay. Tile manufacturers claimed that it was the customers' insatiable demand for red tiles which caused manufacturers to produce a soft red product (or even to sell the under-burnt tiles) without the traditional hard-wearing properties associated with the purple tiles. But the problem may equally have been that manufacturers tried to cut costs by economising on the long drying time normally associated with a tile clay and thus producing an inferior product. Most of the firms had moved over to the semi-plastic process, and invested heavily in machinery.[61]

Archaeological evidence certainly confirms that there were problems. The quantity of wasters rose dramatically in the late nineteenth century, when Jackfield was deluged by great heaps of machine-made roof-tiles, discarded around the kilns. Most of these tiles have been stamped with the maker's name, a practice which dates to the period of the formation of a Broseley roof-tile association in the 1880s. Roof-tiles are a common feature in retaining walls and buildings all around Jackfield, and steps and garden walls around houses are constructed with banks of tiles. Tiles are used to great effect in alternate courses with bricks in a shed to the rear of the Craven Dunnill Works. All this suggests that there was suddenly an excess of roof-tiles, created by either falling prices or falling quality.

Whilst the red clays and the tile clays were often used together in manufacturing, the fireclay operations were generally separate. Fireclays are not found at all in the Upper and Middle Coal Measures, but do occur in the Lower Coal Measures, outcropping in Benthall, Jackfield and Ironbridge. For this reason, fireclay industries are concentrated on the southern side of the river, where these measures are near the surface.

The fireclays were typically high in silica (making them plastic) and alumina (increasing refractory properties), with a small iron content. This meant that they fired to a cream or light pink colour. The main refractory seams were the New Mine Clay and the Ganey Clay. There were also white-burning clays such as the Clunch Clay, the Randle Clay and the Clod Clay, which lacked refractory qualities but which could be used for white brick, sanitary wares and floor-tiles. White clays could also be glazed to produce sinks, or coloured with a wash to produce polychromatic brickwork. In fact, many of the blue bricks used in Ironbridge are not high-fired blue engineering bricks, but white bricks, coloured with a blue exterior.[62]

The best-quality white clays were the Ganey Clays, which outcrop in Benthall and in the Deerleap, and are accessible from shafts elsewhere. At Hilltop Farm in Benthall, just opposite Broseley Wood, is a wonderful clay-mining landscape, where clay has been cut out from the hillside, leaving the marks of old roads, quarries and spoil heaps across the slope.

An equally impressive open clayworking landscape in the same measures was mapped at Caughley (fig. 3.8). There, a survey of Ambrose Gallimore's saggar works, dating to 1754 or earlier, showed an area of depressions which proved to be working on a surface outcrop of the Ganey Clay.[63] This clay would have been excellent for saggar-making, and several pieces of unused saggar were in fact found in the area. What was unexpected was that roof-tiles made of a white clay were also found on the site, the only instance of white roof-tiles yet seen in the area. The clay workings could be shown to cut, and thus postdate, a series of bell pits for mining coal, showing that these were at least pre-1760s in date.

The white clays were also mined from adits to the west of the Iron Bridge, above the site of Burtons White Brick Works. As a form of mining evidence, adits disappear very quickly. At least one of these adits – working as recently as 1953 – could be seen only ten years ago, but has since disappeared. The entrances to two others are, however, visible, as is a brick construction which provided ventilation to the mine. The routes of the tramways by means of which clay was removed from the mines do survive, as do the remains of the inclined plane, taking clay over the railway line to the brickworks itself.

There were two main nineteenth-century brickworks using this white clay – Burtons at Bower Yard, and the white brickworks at Bedlam on the site of the old furnaces. The clay was of such quality that both firms were exporting white clay down the River Severn, as well as producing their own bricks. Burton's white bricks were

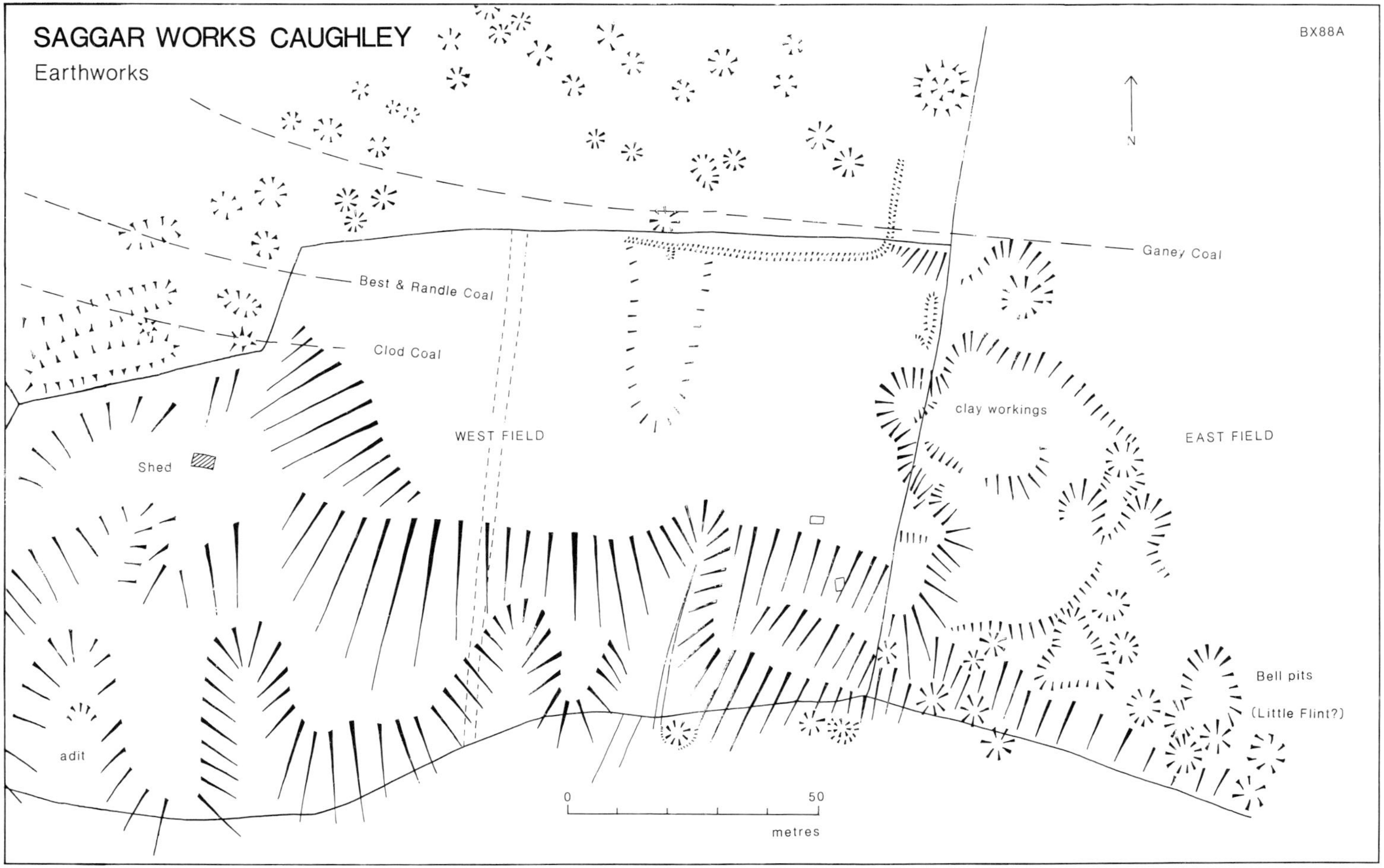

3.8 Bell pits cut by eighteenth-century clay working at the Sagger Works, Caughley (Nicola Smith)

used on the now demolished Ironbridge station, and can be seen in the base of a footbridge which still stands on the site, while bricks from the Bedlam site – stamped 'Madeley Wood Co.' – can be seen on the steps leading up to the church in Ironbridge.

White clays were also used by the earthenware pottery industries, in particular for the manufacture of slipwares. From the mid-eighteenth century onwards these firms were producing press-moulded and wheel-thrown wares with a buff fabric, decorated with a red slip. There were at least three potteries at Benthall, and others at Ladywood, Jackfield, Caughley and Little Wenlock, all of which are located near sources of these white clays. There is field evidence for clayworking at Benthall and Caughley, within a few hundred metres of the pottery site. The distribution of potteries closely reflects the occurrence of white clays on or near the surface.

A local source of white clay was also one of the factors in the development of the clay tobacco-pipe industry in Broseley.[64] The most likely candidate for this clay would be the Big Flint Clay, the uppermost clay surviving in Benthall and Broseley. The clay fired very white, and had excellent plastic properties which would allow it to be moulded into fine clay pipes. It was found in only a few areas, where it was located quite near the surface and could well have been worked out very early; this might explain the concentration of clay pipemaking in Benthall and Broseley near the source of the clay. The shortage of local clay may also explain the transition of the clay tobacco-pipe industry to imported clays in the nineteenth century.[65]

Despite the qualities of local clays, several nineteenth-century firms, including Maws and Craven Dunnill encaustic tile manufacturers and the Coalport porcelain works, were dependent upon imported clays. However, as is so often the case, clay-working practices had changed. Each of these firms had originally been been attracted to the area by the qualities of the local clays, but later changed to imported clays as the demands of the manufacturing process changed.

George and Arthur Maw were already using imported clay from Benthall at their Worcester tile factory when they decided to move their manufacturing operations to Benthall in 1852. George was an amateur geologist, and highly knowledgeable about local clay sources. We do not know precisely which sources of local clay the firm used, but the site they chose at Benthall was located so as to be able to take advantage of both the white and the red clays. In the 1860s, the firm was still using local red clays, as it was considering buying an old pottery at Jackfield for the clays rather than the buildings.[66] Local informants suggest that the firm was using the Randle Clay to make floor-tiles during this century, as well as importing clays.

In fact, one of the Maws' clay-mines still survives. The mine lies to the rear of the original Maws encaustic tile factory site in Benthall. The adit reaches a short way into the hillside, and is lined with plain encaustic tiles, which have very effectively preserved it and prevented the soft sides from collapsing. Small side galleries lead to clayworking faces, left just as they were abandoned, no doubt at the time of the move to Jackfield in 1882.

White clays were the source of a brief revival of ceramic manufacture at Benthall and at the old Bower Yard works in the twentieth century. In response to the massive demand for town drainage, both firms manufactured glazed sanitary pipes – the

Bower Yard works until 1955, and the Benthall site until 1984.[67] One of the reasons for the closure was the popularity of plastic drainpipes. Part of the sanitary pipeworks, including a kiln, still remains within the old pottery works at Benthall, now used for the storage of agricultural machinery. Of the Bower Yard works, nothing remains bar an impressive retaining wall by the riverside built up, in best local tradition, of waste and unusable sanitary pipes.

It would, of course, be wrong to isolate the role of clay from that of coal. Coal and clay were often worked together, particularly by brick- and tile-making firms who used the coal to fuel kilns. Coal, as much as clay, made the Gorge attractive to ceramic industries of all types. However, it is also possible to see a pattern of succession in coal and clayworking. At Jackfield and Blists Hill, shafts which had been sunk in the eighteenth century for coal were in use in the nineteenth century for the extraction of clay. Once again the investment in the landscape generated by the coal industry had been utilised by subsequent industrialists, long after the early mines had stopped working.

The reuse of old coal shafts is also an example of the perception of value in mineral-working. Clay is a low-value mineral, rarely mentioned specifically in mining leases until the mid-nineteenth century. Equally, the sulphur coals were considered of little value and not worked until the nineteenth century, when engine technology made it possible to use what had hitherto been useless coals. Changes in ironworking technology also placed a new value on individual types of ironstone. And the level of perceived value is reflected in turn in the level of investment. The export of coal in the seventeenth century was able to attract a higher relative level of investment than the mining of a small amount of coal by a brickworks in the nineteenth century.

Thus the exploitation of minerals, like any other aspect of the landscape, has to be seen in the context of time and space. It is not possible to rely on recent mining practices to interpret the eighteenth century, because there has been considerable change through time. Mineral exploitation is not just a question of physical supply; the sheer physical amount of each mineral remaining in the ground was often less relevant than the economics of working it. Even today there are supplies of coal and clay, yet underground mining ceased many years ago.

The relationship between raw materials, exploitation and settlement has often been seen in terms of models of economic determinism, implying that natural availability determined the choices made by communities. Yet what can be traced here is a highly complicated pattern in which access to minerals, the individual properties of different minerals, and the level of available investment all overlie the physical geography.

Waste

The relative industrial decline of the late nineteenth century created an atmosphere of desolation which many visitors commented upon. This was created not just by the closing of industries, and the movement of people out of the area, but by the still bare and relatively recent heaps of industrial waste and scars of old mining which dominated the Gorge.

The by-products of industrialisation have done as much as any individual industry to shape the landscape of the Gorge today. Waste, pollution, and the side-effects of old workings, such as landslip, have had a profound effect. When Svedernstierna visited Coalbrookdale in 1802/3, he noted that

> Everywhere in this district one frequently comes across caved-in shafts and hollows in the earth in areas where there are houses and gardens. Some houses had recently collapsed and others were standing at such an angle that I don't understand how people could live in them.[68]

Few of the contours we see are those of the original landscape, now obscured by heaps of roof-tile waste and mountains of coal- and clay-mining debris. No garden is without its complement of pottery; roads are made of slag or pipe waste, and boundary walls of old sanitary pipes. The pattern of present-day building, of woodland, of agriculture and even of settlement as a whole can all be related to the lasting effects of past industries. Waste is another key to reading the landscape.

The most obvious category of waste is that from iron-smelting and founding. In the blast furnace, impurities in the iron-ore and coke combine with limestone to form a mass which floats on top of the pure iron. This is tapped periodically, and on cooling forms a glassy slag. The local cold blast-furnace slag of the mid-nineteenth century is a marbled light and dark blue material, with few bubbles, whereas that from the eighteenth-century Benthall furnaces is black. Old iron was always recycled in the blast furnace, and so practically never survives in any quantity on archaeological sites. Equally, any slag with a high iron content, such as that made in the rather inefficient early charcoal furnaces, would be recycled.[69] But ironworks produced many tons of unusable slag each year, as well as old casting sand, boiler ash from steam engines, and slags from the activities associated with ironworking – forging and casting.

Removing this waste from ironworks was a constant problem. Cinder heaps were frequent landmarks in the past, surviving today only as field names. Once water-power was replaced by steam, old pools provided a useful dumping ground for slag, as at Horsehay in the nineteenth century, or Coalbrookdale today. Slag occasionally appears as a building material, topping boundary walls at Coalbrookdale, making up the splendid retaining wall which was recently demolished at Horsehay, or dumped by the roadside in great blocks, as at Coalport. Otherwise, it has simply built up on furnace sites in great quantities.

At the Benthall furnaces, situated in a narrow valley, removal of slag was a constant problem. The owners of the furnaces were accused of dumping cinder and ashes on the road in 1781, and in 1795 were obliged to pay their landlord for the right to dump waste. One solution they adopted was to use the slag and ash for surfacing roads; they contracted with the proprietors of the Iron Bridge to surface Bridge Bank, the Iron Bridge itself and parts of the Wharfage with 'three tons of furnace cinders to each yard in length of a road nine yards wide, and thirty loads of ash to cover them'.[70] This ash can often be seen in modern service trenches.

The use of slag for road surfacing became an important industry in its own right after the First World War, when two firms operated at Blists Hill between 1920 and 1941, setting up crushing equipment close to the old blast furnaces.[71] The crushed slag was mixed with tar, and then sold commercially for road-making. Although nothing remains of these operations, they have left a legacy of instability on the site today, caused by digging out and redumping great areas of fill.

Ironworking must also have affected agriculture, or at least the value of agricultural land. A local land surveyor commented that at Coneybury furnaces the land was 'chiefly mounted upon and entirely spoiled for agricultural purposes', whilst there was a field near the Broseley furnaces which 'when in grass is certainly at times injured by the Smoke from the Furnace'.[72]

The damage from coal-mining was much more widespread than that from ironworking. It was the fields in and around Broseley and the slopes leading down to Jackfield which were most 'trespassed upon by the working of the mines'. Admittedly, the tenant or miner was obliged to pay damages to the landlord during the course of working, and to level the mounts and make good the land at the end. Nevertheless, it was Vickers' opinion that at the end of working most such land would be worth relatively little in agricultural terms,[73] and although the landlord might gain something by this arrangement, tenants of neighbouring houses often suffered. Occasionally a tenant was obliged to replace a house spoiled by mining, but more commonly houses seem to have been allowed to collapse.

Coal-mining, in conjunction with ironworking, was in fact responsible for the disappearance of several settlements (fig. 3.9). Groups of houses formerly existed at Mone Wood and Coneybury, near the furnaces, and at Croppers Holes just below the Benthall furnaces. By the late eighteenth century, several of these houses had been rendered uninhabitable by the pits.[74] Today there is no settlement at Coneybury, and only one or two cottages in what were formerly Croppers Holes and Mone Wood, although remains of houses could still be seen in the early 1960s. Settlement, therefore, was as well able to contract as to expand.

The pattern of woodland today is also related to early coal-mining. Very dense areas of early bell pits at Deerleap, at Caughley and at The Mines have remained as woodland, as the spoil heaped onto the surface rendered the land useless for agriculture. The deeper shafts of the nineteenth century left great standing pit mounds which still tower over Broseley today. These mounds, too, tend to be covered with trees which provide some measure of stability to their steep sides. This is one of the reasons why it is in the woodlands that some of the best-preserved archaeology survives – mining made it unusable for agriculture, and thus preserved it from the problems of deep ploughing which have subsequently destroyed many sites.

By far the most damaging legacy of mining has been surface landslips, created by underground voids. Miners were not always scrupulous about respecting surface boundaries, and several Broseley buildings have been affected. The Red Church at Jackfield was reputed to be haunted by a ghost who threw down the partly completed building. The church suffered constant problems and was later demolished. The culprit was said to be nearby clay-mining, which had undermined the foundations of the building.

3.9 The after-effects of the Jackfield slip (IGMT1982.240)

During the equally dramatic Jackfield slip, houses moved up to 18 m, and roads split open.[75] As a result, yet another community – the Salthouses – was totally demolished. The problem lay with the Wallace pit just above the area, dating back to at least the 1760s, and mined first for coal, and later for tile clay. The pits were pumped during working, and during World War II when the site was used as an ammunition dump, but after the war pumping ceased, and the ground surface sheared away from the more solid bedrock. Slipping continues in the area even today, with so much movement that only a temporary road surface can be laid across the site. It is certainly unsafe to build in this area today.

Slips also occurred on the north bank, in the Lloyds area, although it is less easy to attribute this directly to mining. Again, all of the documented cottages have gone, and the effects of slip are such that the Lloyds is one of the few areas where none of the eighteenth-century plots can be identified.

Ceramic industries, too, left an imprint on the landscape. Like blue slag, pottery waste is found as surface makeup on local roads, frequently some distance from its likely origin. But its impact on the landscape is tiny in comparison with the havoc wrought by the debris of roof-tile manufacture. With the changeover to more mechanised production after the 1870s, the quality of Broseley tiles seems to have fallen, and huge numbers of cracked or unusable waste tiles were produced. Some were used in novel ways as steps, boundary walls or even instead of bricks in building construction, but most were dumped around the works, totally burying the site of the old Calcutts furnaces, or spilling down onto the side of the railway below Hollygrove. A very large proportion was dumped downhill onto the river bank at Jackfield, where erosion has exposed several metres of unstable tile waste. The act of dumping is an indication that the riverside and towpath at Jackfield had gone out of use by the late nineteenth century, and that there was relatively little civic pride or planning control operating in the community.

Dumped waste material often created usable land for industries which were short of space on narrow or unstable sites. At Coalport, excavations have shown the southern ranges of the works to have been built on top of 2–3 m of tipped porcelain wasters, lying on the river banks.[76] This was no doubt a response to the floods of 1795 which caused some damage to the riverside site. At Maws, too, on the south bank of the river, masses of tile waste have been dumped on the bank. The resulting platform created extra land, used eventually for additional buildings, and even a tennis-court. Even the original Maws site at Benthall is now a platform, several metres above the valley floor, where a combination of ironworking waste and tile waste has been used to create additional flat land. Industrial waste was thus a convenient and cheap form of landfill.

The influence of waste on the landscape should therefore not be underestimated. It has been demonstrated that waste and old workings clearly affect the pattern of settlement which has survived today. Agriculture is limited, partly by the lack of flat land, but also through the effects of coal and iron waste on soils. New building today is difficult in areas of old mining, areas of landslip, and on parts of Jackfield heaped with old tile waste. And finally, even the shape and form of much of the landscape today is not natural: it represents the contours of the aftermath of mining, and of the waste products which industry has created.

The landscape of the Gorge today, therefore, can be seen to be a direct product of centuries of mining. Housing, agriculture and industry have all been influenced not only by the availability of minerals, but by the legacy of past mining practices. This is a clear example of the need to see the minerals not just as a resource, but as a determinant, whose effects upon the landscape need to be understood through time and space.

4

THE HUMAN LANDSCAPE

There is no doubt that the physical geography of the Gorge had an impact – whether advantageous or detrimental – on the pattern of industrialisation. The steep-sided valley with minerals outcropping on each side, the small but fast-flowing streams and the river at the bottom provided a physical framework which shaped settlement and industrial development. Given that landscape evidence provides only part of the wider industrial picture, we have nevertheless shown that it is impossible to ignore the physical setting of the Gorge in understanding the decisions made by industrialists.

But to see industrialisation in the landscape just in terms of the effect of natural resources on the activities of industrialists is too simple a model. Superimposed upon the natural landscape is what could be described as the human landscape – a landscape constantly modified and changed by human activity. The human landscape is made up of all the landscape features created by past occupants – transport links, pools, housing, mines and even piles of industrial waste. The very existence of these features dictates where and how later generations can operate. In this way the landscape becomes not just a resource, but a determinant, playing a significant role in defining the options available to an industrialist at any given moment. It is vital to consider this inherited, human landscape as well as the natural landscape in any analysis of the field evidence for industrialisation.

One of the concepts which can be found within this human landscape is evidence for technical change. The adoption of innovations can be seen clearly in the layout and technology of industrial sites, as well as in the choice of their location. As elsewhere, the evidence from the Gorge shows that technical progress is neither simple nor linear, nor does it occur in isolation from wider economic considerations. The influence of individual engineers cannot be studied here, but it is possible to see the take-up of innovations in the context of the historical landscape, and to explore the relationship between the two.

This chapter therefore takes a systemic approach to the field evidence, looking at the human landscape and the light it throws on industrialisation. In order to do this, two aspects of industrialisation have been chosen which can be readily seen on the ground – power and communication.

Landscape of power

Minerals were not the only factor which attracted the early ironworking industry to the Gorge: the production of iron in a blast furnace required large amounts of air to be blown into the furnace which in turn depended upon a form of motive power. The source of this power was water. Among the most important resources to be found in the Gorge in the eighteenth century were the streams which rushed down the steep sides of the valley into the River Severn. Water and height provided the power for ironworking (fig. 4.1).

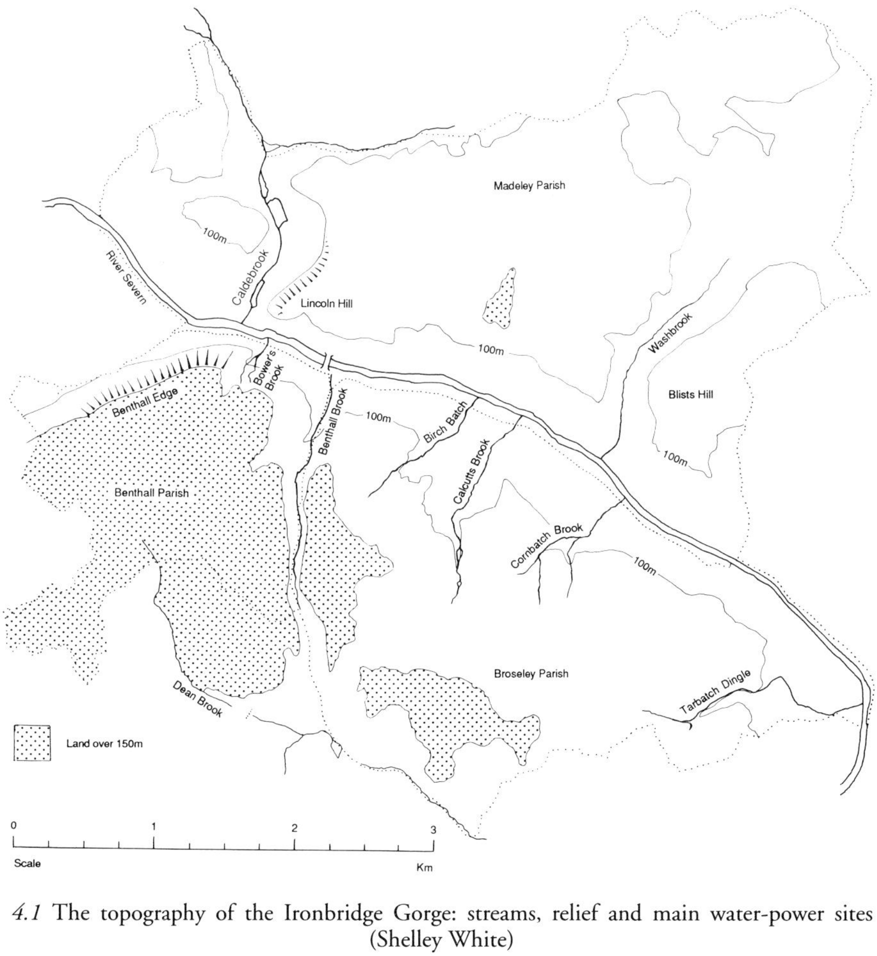

4.1 The topography of the Ironbridge Gorge: streams, relief and main water-power sites (Shelley White)

Unlike minerals, this resource could not be transported, so ironworks were, until the development of steam-powered blowing, invariably located on streams or rivers. Yet even in the eighteenth century it is possible to see important changes in the choices ironmasters made about the location of their works. These decisions are a useful source of evidence for the way in which innovations were adopted during one of the most crucial periods in the development of ironworking technology – the 1750s.

When Abraham Darby came to the Gorge in 1709, he found an existing ironworks consisting of a furnace and several forges with a complex water-power system.[1] At the top of the Coalbrookdale Valley, a seventeenth-century charcoal-fired furnace was built into a bank, formed by the dam of a pool. The pool was located at the junction of two streams, one from the north-east, and one from the north-west. Of all the valleys feeding into the Gorge, this was the only one with two well-established streams and a good supply of water from a large catchment area. There was also the advantage that the natural topography sloped steeply at this point, and so there was plenty of height to allow an overshot wheel to operate.[2]

The Upper Furnace Pool was not the only pool in the valley. Further down the valley at Coalbrookdale another pool had been created to power the Great Forge, a pool which could well date back to the earliest ironworking in Coalbrookdale in 1536. Below this was what was later known as the Boring Mill Pool, just above the Middle Forge, and at the bottom of the Dale the Lower Forge Pool, powering the Plate Forge. At the top of the Dale, above the Upper Furnace Pool was a new pool, built perhaps by Shadrach Fox in 1698, not to power any ironworking directly, but to act as a storage reservoir for the Upper Pool. At least one of the Coalbrookdale pools originated as a pool for the mill of 1520, a pattern which was repeated commonly in the choice of an ironworking site.[3] Thus, Abraham Darby inherited a well-established water-power system, parts of which dated back to the sixteenth century. In itself, this water-power system already represented a considerable financial investment, to which in 1715 he added a sixth pool to power his new Lower Furnace, immediately below the Upper Furnace (figs 4.2 and 4.3).

But by 1734 it became apparent that there was insufficient rainfall during dry summers to keep both furnaces in full operation.[4] In order to augment the system, a horse gin was installed, to recycle water through the system. The gin was by the new furnace and therefore must have lifted water about 10 m from the Upper Forge Pool back into the Lower Furnace Pool. The company could have chosen to use one of the steam engines they were currently manufacturing to do the job, but had found, on comparison, that the horse gin was more efficient than the steam engine. This arrangement worked well, and for the first time was able to get iron production up to the level of a charcoal blast furnace.

The horse gin had always been seen as a temporary measure, and by 1744 it was replaced by a Newcomen fire engine with a greater capacity. The engine seems to have been located on the site of the present Coalbrookdale Chapel.[5] There is no direct evidence for the working of the pump, however; the engine site is about 10 feet (3 m) above the Upper Furnace Pool (242′ OD). Water could therefore have been lifted into a lade, and run back down into the pool to serve both the Upper and Lower Furnaces (fig. 4.4).

This engine – and its horse-powered predecessor – established the fundamental principle of recycling water to keep a blast furnace in continuous production. A blast

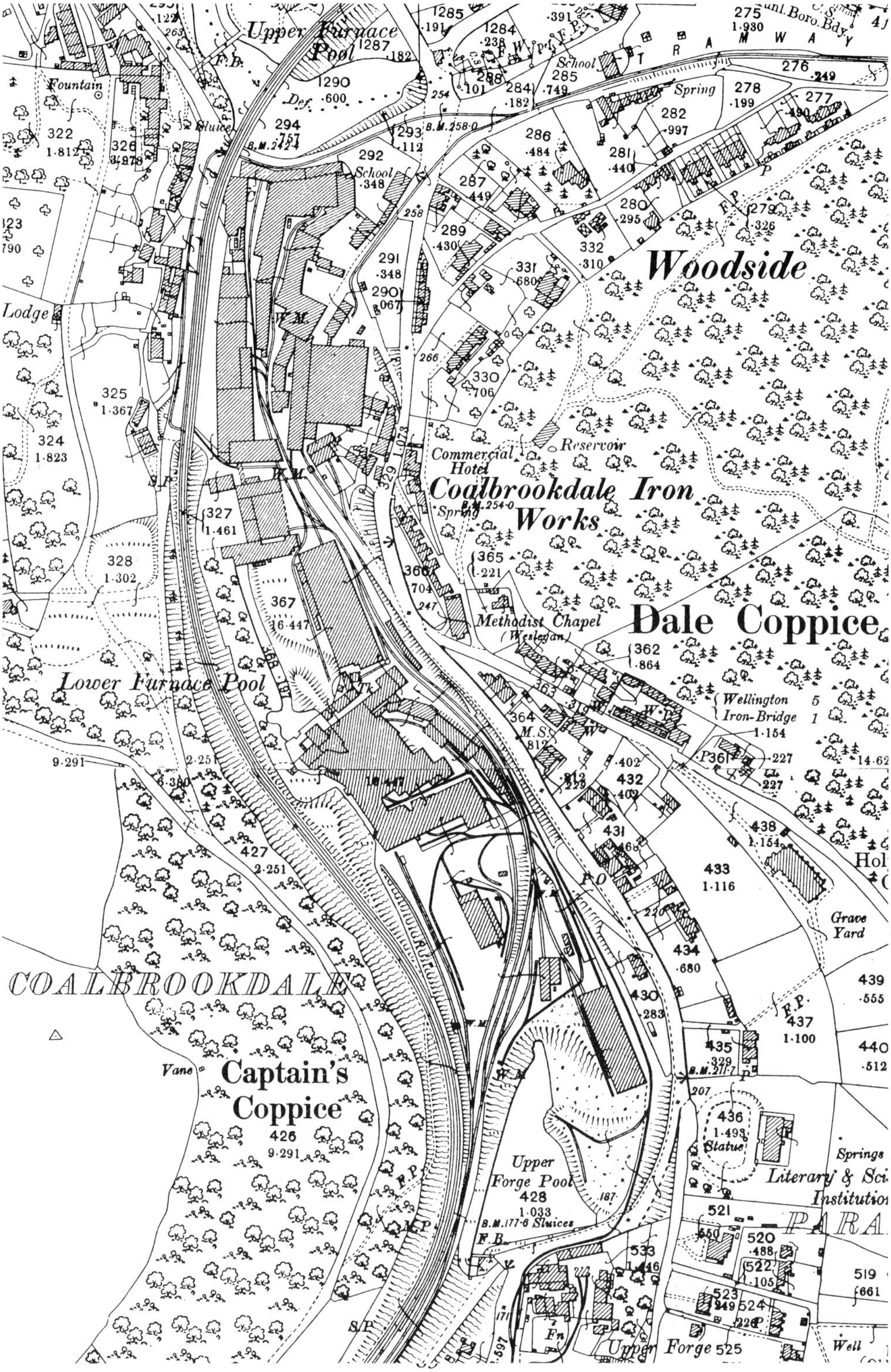

4.2 'The 1902 Ordnance Survey Map for Coalbrookdale, the map edition on which the Nuffield Survey was based. This section shows the Coalbrookdale Works and its associated pools.

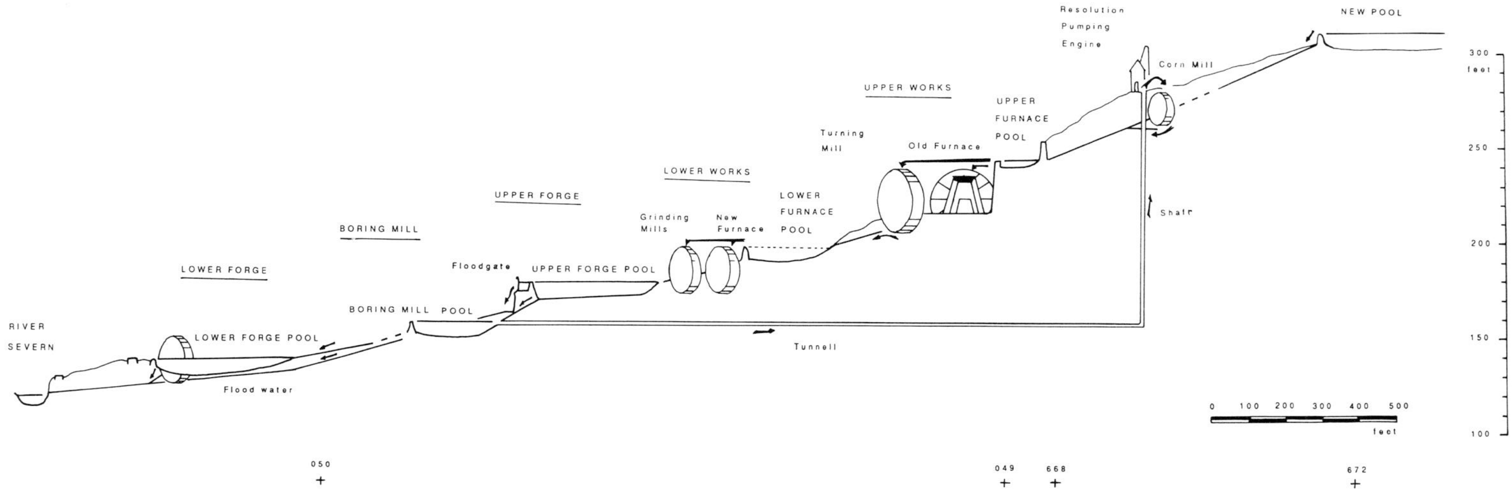

4.3 A schematic section through the Coalbrookdale Valley showing water-power installations, *c.* 1801

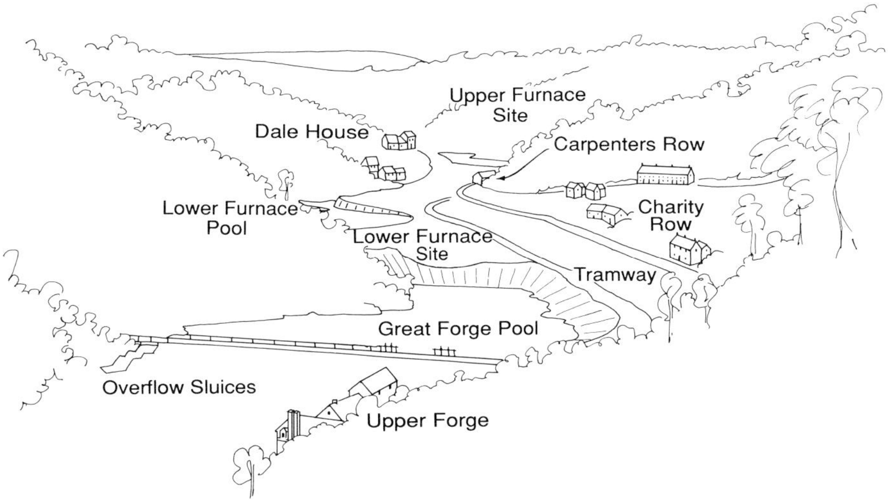

4.4 The system of pools and ironworks in Coalbrookdale, based on an illustration by J. Farrington, 1789

furnace did not necessarily require a very large amount of water, but steady and uninterrupted blowing, and ironmasters could not risk the supply of water running out; mechanical pumping proved the solution, and had a demonstrable effect on increasing output.

The principle of recycling water was adopted from the outset at each of the new furnaces built in the area as part of the extraordinary expansion of ironworking during the 1750s. It is very clearly reflected in the change in the type of sites used for ironworks. Previously the emphasis had been on achieving a fast-flowing, reliable head of water with as much height as possible, through dams or long leats. Now it was necessary to have a steady supply of water, but height and volume were no longer so important.

The first of the ironworks to employ this new principle in its construction was Horsehay, built by 1755.[6] The partners chose a traditional site below an old mill pool, which they had (with some difficulty) to reconstruct. This pool provided very little height and was only fed by one small stream. Several other pools were also built – one to the north-east as a holding pool, and two pools below the works, fed by a different stream. Water ran from these lower pools through a tunnel to the bottom of a pumping shaft, where it was lifted directly over the water-wheels. It seems, therefore, that the large Horsehay Pool acted mainly as a storage reservoir, and was also useful for cooling. It had no direct role to play in providing power.

The idea that a water supply could be below the ironworks is nowhere better demonstrated than at the Madeley Wood Ironworks on the banks of the River Severn, in blast by 1756.[7] The furnaces there are built into the hillside, to give the traditional advantage of high charging; but, unlike any earlier furnaces, there was no water-supply above the works. Instead, water was pumped up from the river and directly over the water-wheel (or via a small holding pool). None of the investment in dams or pools was necessary, the fluctuating levels of the river were no problem[8] and there was a constant supply of water.

This location pattern was repeated at the sites chosen for the New Willey Ironworks (1757), the Lightmoor furnaces (1756), at Ketley (1778/9) and at the forges at Wren's Nest (1779). Each was located on a small stream with relatively little height, sites that could not have been contemplated without the capacity for recycling. Nevertheless, each site still required land and capital investment in pools for holding water – there were four pools at the Willey furnaces and three at Ketley – and ironworks were still irrevocably tied to a water supply.

Logically, direct blowing should have liberated ironworks from dependence on all but the most minor supplies of water, and enabled them to be located much closer to supplies of raw materials. Direct blowing was the next major technical innovation to affect iron-smelting. The combination of cylinder bellows, patented by Isaac Wilkinson, and the new Boulton & Watt engines was first successfully tried by John Wilkinson at the New Willey site in 1776. Yet, as is so often the case with innovations, direct blowing was not adopted immediately or, for some time, universally. In order to understand this it is necessary to look at the landscape evidence.

The initial advantages of the Coalbrookdale Valley, with its string of pools, soon proved to be a very real restriction. Having established a system of water recycling, the Coalbrookdale Company was reluctant to replace the existing water-driven blowing equipment and wheels. Thus, faced with further water shortages in the 1770s, the company installed what was then one of the largest steam engines built, the *Resolution*, to extend the recycling system. Water was brought from the Boring Mill Pool, through a tunnel half a mile long, up a shaft 120 feet (37 m) high and back down to the Upper Furnace Pool. Rather than invest in direct blowing, as they did at other sites, the company chose to continue to adapt their working methods to the restrictions of Coalbrookdale. Ironically, the result was a situation in which the 'most work [was] done at these places with the least water, of any place in England'.[9]

With the exception of Coalbrookdale, direct blowing was almost universally established by 1802. Some companies were able to adapt their recycling engines located (unlike Coalbrookdale) close to the furnaces, and others brought in new engines.[10] New ironworks which incorporated the principle directly could be located away from streams and, as a result, there was a migration of furnaces to the north, closer to new supplies of coal and ironstone and to transport networks.[11]

Forging and many of the other activities associated with shaping iron were also dependent upon water-power – one of the factors which long acted against the integration of furnaces and forges, as few sites had the capacity to supply both .[12] The new blast furnaces sent their iron off to distant water- (and charcoal-)powered forges and it was only in the 1780s, when steam-power was applied to hammering, rolling, slitting and blowing, that smelting and forging activities were once again integrated.

Water also remained necessary for cooling – the new rolling mills at Horsehay had their own water outlet, and relied on the large surface area of the pool to cool and recycle the large amounts of water needed to cool the rollers.

The continued importance of water-power to the iron industry even after the introduction of steam is apparent from landscapes such as the Benthall Valley (fig. 4.5a–b).[13] The narrow Benthall Valley falls over 100 m in a distance of just over a kilometre. It rises in the area of the Deerleap, but is today culverted down most of the valley. In places the brook is supplemented by runoff, but otherwise the catchment area of the stream is small. In the valley, there were originally three mills and their pools, and pools for a boring-mill, machine house and the furnaces. Unlike Coalbrookdale, the valley is narrow, and thus any water-powered equipment would have had to make use of height rather than volume to generate power. The lowest corn-mill was famous for its high, narrow water-wheel, and there was also a boring mill, described by the engineer John Rennie.[14]

The pools at Benthall were still in use in 1835, and indeed a water-wheel still operated at Coalbrookdale in the 1920s.[15] A small water-wheel base survives at the Coalbrookdale Lower Forge site today, and was used to generate electricity in the 1930s. It is very apparent that, despite the efficiency of steam as a source of power, there were situations where it was not used. There may have been many reasons for this, including the prohibitive costs of the royalties payable on Boulton & Watt engines, but equally one has to consider the natural advantages of the landscape of the Gorge, which provided free power through the fortuitous combination of water and height.

Water-power was one of the resources which attracted industry to the Gorge in the early eighteenth century. The principle of recycling water, developed as a solution to the problems of Coalbrookdale, was a key factor in the rapid expansion of ironworking in the 1750s. Ironically, the very efficiency of this system, and the cheapness of water-power, militated against the adoption of technological innovations such as direct blowing later in the century. It is clear once again that the adoption of technology takes place not in a vacuum, but in the context of many different factors, including the existing, inherited industrial landscape.

Communication in the landscape

The natural form of the Gorge with its steep and unstable sides, dropping over 100 m to the river, inevitably created transport problems for industrialists. Access to the riverside was difficult down the slopes, and crossing the river could be dangerous. There was no large resident population in the area, so goods had to be transported away from the Gorge to markets. Equally, as supplies of locally occurring minerals diminished, they had to be brought into the area from outside. The whole pattern of industrialisation through time has to be seen in the light of transportation, and how industrialists dealt with the problem of getting in and out of, as well as across, the Gorge.

The paucity of known medieval, Saxon or earlier settlement in the Gorge is evidence in itself for the problems of access. Apart from a Bronze Age find in the

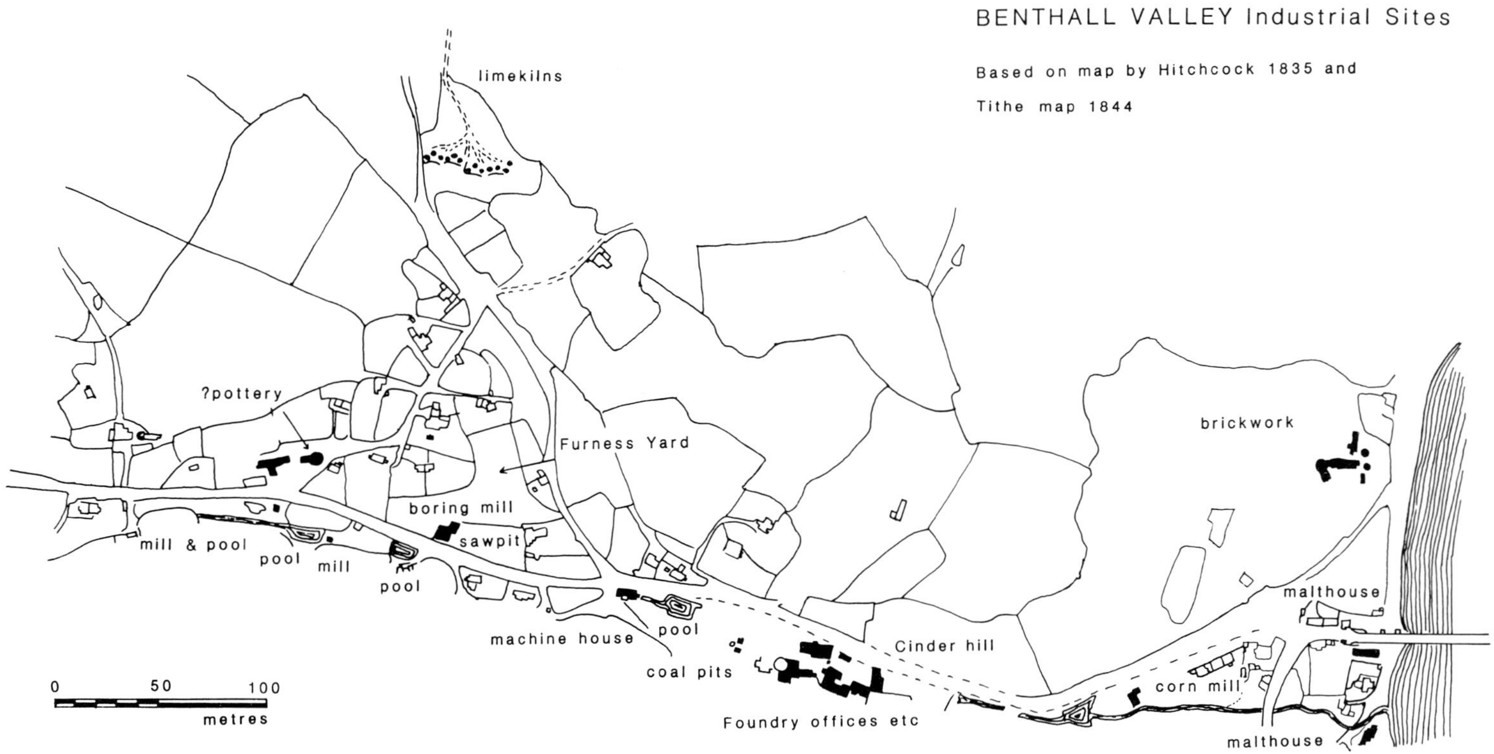
BENTHALL VALLEY Industrial Sites
Based on map by Hitchcock 1835 and
Tithe map 1844
limekilns
?pottery
Furness Yard
boring mill
sawpit
mill & pool
pool
mill
pool
machine house
pool
coal pits
Foundry offices etc
Cinder hill
corn mill
malthouse
malthouse
brickwork
0
50
100
metres

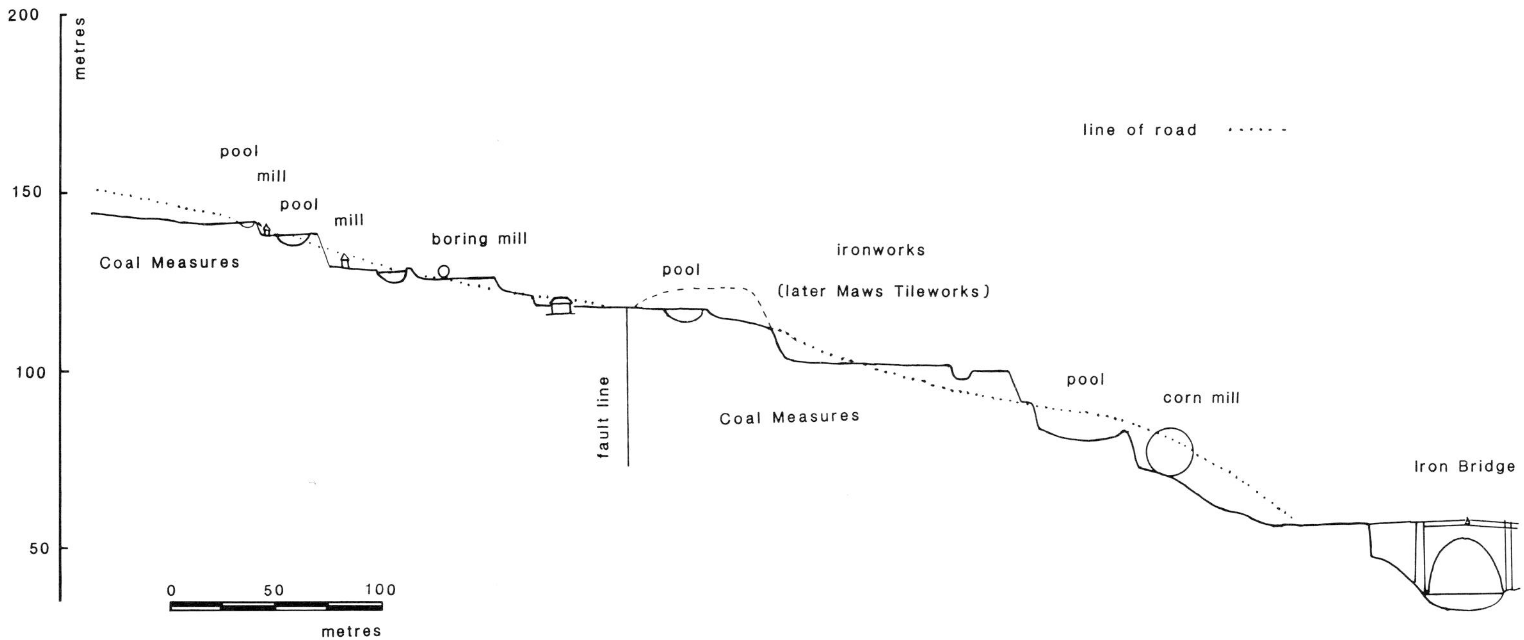

4.5 The Benthall Valley, like Coalbrookdale, included a series of pools and industrial sites: (a) map (b) section

river, there is nothing so far to suggest prehistoric occupation. There was very little medieval or earlier settlement in the valley, and the earliest known roads bypass the Gorge well to the north or the south. River crossing-points are often natural foci for settlement, but the nearest major crossing and settlement is at Bridgnorth.

The tramway network

It was only with the growth of the coal industry in the seventeenth century that there emerged a need for good access to the banks of the Severn. Easily accessible coal seams and the value of the River Severn as a means of transport to markets were two important factors in the growth of the Severnside coal industry. Unfortunately, many mines were located in Broseley, or on estates such as John Weld's, well away from the riverside. Coal was bulky, and the existing road network minimal. An early solution to the transportation of coal was the development of the wooden tramway.

Broseley is well known as one of the earliest places in Britain to have used tramways.[16] In 1608, Star Chamber proceedings record a dispute between Broseley landowner James Clifford and his tenant, Richard Wilcox. Wilcox agreed to allow Clifford to dig pits for coal on the land he held as tenant, but after a dispute over compensation due to Wilcox the latter was given the right to dig his own pits elsewhere. The dispute erupted into violence, with miners imprisoned down shafts and cottages demolished. As part of the proceedings, there are detailed descriptions of 'a vary artificiall Engine or Instruments of Timber . . . therewith to convey coles . . . unto the banck of the River of Severn', which has been interpreted as an early example of a wooden railway.[17]

Wooden tramways continued to be used by the local coal industry and, from the 1740s, were adopted by the ironworks. It was as a result of this association with the coal industry that the expedient of using iron rather than wood for the rails was first adopted at Coalbrookdale in 1767.[18]

Archaeological evidence for both wooden and iron tramways is fragmentary. Part of the problem is locating the actual routes. Where they are documented, for example in mining leases, it may be possible to identify the route through field names. But, as William Ferriday remarked in 1755, many tramways were 'of the mushroom breed, only the produce of one night'.[19] The rails would be taken up at the expiry of a lease and most likely reused elsewhere, so there would be very little remaining on the ground. For example, a ploughed field at Rowton through which a railway was known to have run produced no evidence for a change in soil colour or a dip in the contours which might suggest a tramway route. At ironworks, old rails, like any other scrap iron, would have been broken up and remelted in the furnaces.

Continuity with later routes would logically be another means of identifying tramway lines. Yet if this is assumed within the small area of Jackfield, there are at least four different contenders for the most likely route of James Clifford's railway line.[20] In the few instances where tramways are mapped, as at Blists Hill in 1840, 1886, 1902 and 1927, the rate of change over 20 years is enormous, suggesting that tramways were still being laid in a temporary fashion to serve changing industrial needs.

This phenomenon explains why some tramway routes do not conform to the image of the nineteenth-century railway, carefully laid level in cuttings or embankments and often taking a longer route to avoid climbing. Fieldwork suggests that the majority of tramway routes did not create the hollow ways one would expect from roads, and involved little investment in earthworks. Many were not at all level – for example routes through Caughley climbed and dipped over rather than went around hills, and engravings show horses hauling cylinders up the steep slope of Jigger's Bank in Coalbrookdale.

Nevertheless, some clear routes can be identified and are still in use as footpaths. Those in Jackfield, and in the woods to the north of the Tuckies are well laid, level and about a metre wide. The Gorge is full of footpaths, which most likely originated as tramways. For example, a track hugs the eastern side of Coalbrookdale, passing below the wall of the church. A plan of 1788 to build a canal from the foot of Brierly Hill, along the side of the Dale, was abandoned, and a railway built instead on the same route. The canal route must have been level, and investigation shows that the Coalbrookdale route follows the 275-foot contour, at precisely the height of the foot of Brierly Hill. This evidence does suggest that, by the later eighteenth century, tramway routes had become much more permanent, and involved greater engineering input than earlier lines.

One of the few surviving tramway bridges can be seen at Newdale – a stone structure crossing a stream. Although it has not been documented, the bridge is directly on the line of an excavated tramway, and almost certainly must have carried rails.

There is some archaeological evidence for the rails themselves. Wooden rails have been found by chance in an excavation to the west of Bedlam Furnaces (fig. 4.6).[21] Oak sleepers were used to space lengths of oak rail, set on a bed of furnace slag and occasional bricks. The rails themselves were not flanged, but the one surviving wooden railway wheel has a thick, exterior flange. The gauge of the rails was 1.14 m (3 ft 9 in) – narrower than northern gauges, but typical of Shropshire. One wooden wheel has also been found, with a flange suggesting that it was a tramway wheel.

The transition from wood to iron in tramways was gradual. Flanged iron wheels were first adopted, followed by iron rails.[22] Only later were flanged rails taken up. The first rails are known from historical accounts, and consisted of iron plates 6 feet long, set on wooden bases (a possible alternative explanation for the Bedlam timbers). No examples of these were thought to have survived until work at Newdale identified pieces of iron set into a reconstruction of the north building, dated to 1768. The pieces of iron were broken in two, with holes for bolts at either end, and match closely historical illustrations. These could well be the first examples of Coalbrookdale rail.[23]

Sections of flanged iron plateway occur more commonly around the area, often reused in buildings as lintels. *In situ* plateway has been found at Coalbrookdale behind

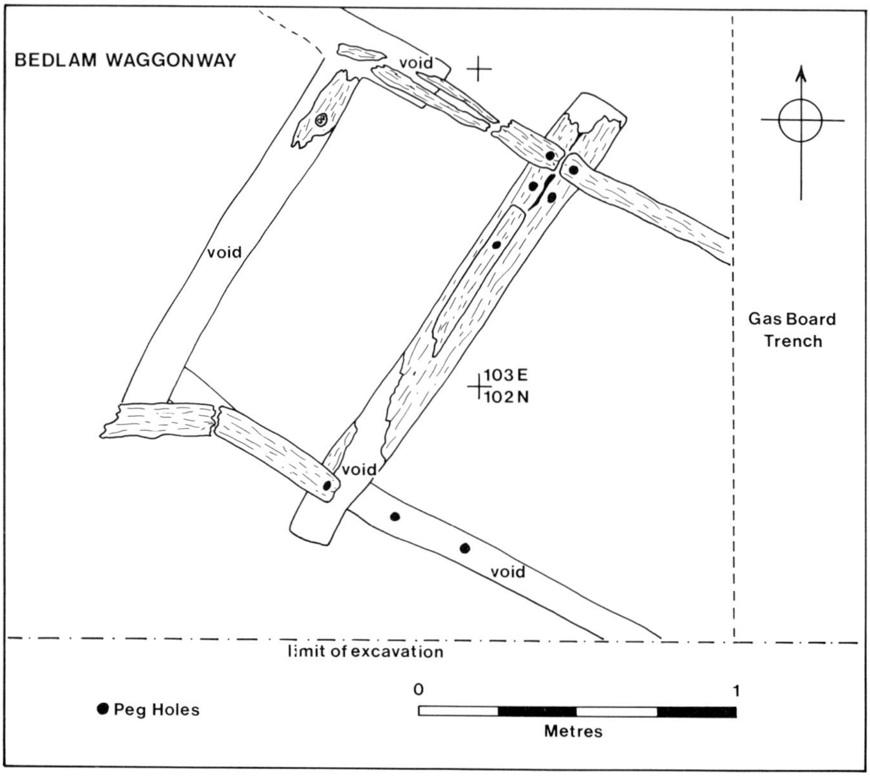

4.6 The Bedlam wooden tramway

Rose Cottage, and also at Blists Hill associated with the Heslop engine base. The rails are of cast- or wrought-iron, flanged on the inside and bedded in ash or slag.[24]

Even after the construction of the railways in the 1860s, horse-drawn tramways (or plateways) remained in operation for local use at ironworks, continuing until well into this century. Maps show that, on sites such as Blists Hill, there was a far greater density of tramways in the late nineteenth century than can be documented for any other period. Tramways continued in use around Coalbrookdale and at Horsehay as local transport networks, often with interchanges to the main railway network (fig. 4.7).

The use of tramways provides another example of a local innovation of early date, which remained in use for three centuries as a perfectly serviceable, if somewhat antiquated, practice. However, although tramways provided a solution to moving goods along bad roads, there remained the problem of negotiating the steep slopes to reach the riverside. Even where coal-mines opened out onto the hillside, their entrances were usually some way above the river itself, and few routes around the Gorge could be devised which did not involve a slope too steep for a horse and cart. The solution was the use of the short incline or 'wind'.

4.7 A team of horses brake a loaded coal wagon as it descends Jigger's Bank, Coalbrookdale. William Williams, 1777 (Shrewsbury Museum)

The principle of the incline was basic:

> a large barrel or wind is fixt at the Top on which runs a large Chain; and at the end of this is a Wooden Waggon that will hold 2 tons each. This chain reaches to the river, and when the Waggon is loaded at the Top it gets a going gradually, which brings the empty one up, and so continues till the vessel is laden.[25]

Several inclines are known from documents. Names such as 'Wynd' on a cottage at Jackfield refer to the sites of wind-houses for old inclines. The earliest device of this type is James Clifford's 'tylting rails' of 1607 – interpreted as a short, rope-worked incline near the river. Such inclines were later used at coal-mines in Madeley Wood, the Tuckies, the Lloyds and probably in Ladywood to let coal down to waiting boats. Further away from the river, there was a tramway incline on Jigger's Bank which until 1802 served the Coalbrookdale ironworks.

Fieldwork has also revealed the lines of several other tramways. By far the steepest and longest slope to be negotiated is the drop from the quarries at the top of Benthall Edge down to the riverside. There is a series of inclines down which limestone was lowered, one of them clearly replacing an earlier, rather tortuous zigzag path up the hillside. An incline relating to a documented eighteenth-century tramway route is also the most likely explanation for the massive clay ramp, several metres wide, to be found in the woods below Rowton.[26]

The combination of inclines and tramways thus provided an industrial transport network for moving bulky materials around the Gorge. However, the principles established by the builders of the tramway network were also applied to other forms of transport such as the canal. In fact, it is unlikely that a canal link between the Severn Gorge and the industrial areas to the north would ever have been contemplated without the inclined plane having first been developed as a means of descending into the Gorge.

The canal network

Canals, like tramways, were restricted by the awkward topography of the Gorge. The slope was too steep, and the quantity of water required too great, to contemplate a flight of locks as a means of getting from the top of the slope to the bottom. The principle of the inclined plane, as developed on the railways, was therefore used. It was first applied to canal construction at Ketley, about 4 miles (6 km) to the north. There, instead of letting railway wagons down the slope, tub boats were floated onto a cradle, and lowered down two parallel sets of rails to a canal basin at the bottom.[27]

Two short stretches of water-filled canal at Coalport and Blists Hill, linked by the Hay Inclined Plane, are all that remains of the Shropshire Canal network in this area.[28] In 1788, it was proposed to build a canal, to run south from the existing canal at Oakengates, dividing into two branches linking with the River Severn at Coalport and Coalbrookdale.[29] The problem was that each branch terminated at a point at least 100 m above the River Severn.

Under the original scheme, the western branch of the canal was to terminate at Brierly Hill just above Coalbrookdale. From here, goods would drop down a pair of shafts to another canal branch, running along the eastern side of the Dale. An incline would then take goods down to the riverside.

We know that the Brierly Hill shaft system was built as designed.[30] The canal divided into four fingers, with two 120-foot-deep vertical shafts between them and a winding-drum between the centre pair. Goods were to be wound down the shafts into trucks on a short railway line leading out through a tunnel. One of the shafts and part of the wall has been exposed, and it has been possible to identify the tunnel opening out of which the tramway came. From there, the tramway led down and across the valley on the dam of the Upper Pool, linking with a further tramway route.

Originally it was proposed to build a second stretch of canal along the valley side. A puzzling 'navigation' appears on early maps,[31] but otherwise there is no evidence for the canal. However, there is a clear level route, following the contour around the side of the valley and marked by occasional stretches of stone wall, and slightly staggered field boundaries. Thus field evidence conclusively shows that the tramway was built. Less easy to identify is the inclined plane at the bottom of the Dale.[32]

The shafts at Brierly Hill are cut by the deep swathe of an inclined plane leading south, which seems to support the documentary evidence for the replacement of the system in 1794. Archaeological survey, however, has revealed a more complex picture.[33] The first inclined plane to be built in fact lay to the east, and was linked

4.8 The Hay Inclined Plane while still in operation in the 1890s (IGMT1982.2713)

to the existing tramway route across the dam. On stratigraphic evidence, the second incline could only have been built once the canal system went out of use. The second incline links with a different tramway, leading to the Coalbrookdale works. It seems that the shafts were in operation until 1794, when they were replaced by an inclined plane. A second inclined plane was built later, probably relating to the permission given in 1800 to the Coalbrookdale Company to build a tramway on the canal towpath leading to the Horsehay and replacing the old Jigger's Bank line.[34]

This complex sequence of events makes sense of a fundamental discrepancy between the documentary evidence and what could originally be seen on the ground – the six years from 1794 to 1800 during which both the canal and an inclined plane operated. It also underlines the difficulties of getting goods in and out of the Gorge, seen in the large number of short-term and rather awkward solutions.

The same problems were faced by the builders of the eastern branch, terminating above Coalport, but their solution was more effective and longer-lived. Here, an incline was to take goods down 207 feet to a stretch of canal parallel to the river (fig. 4.8). Because the river level fluctuated, there was no lock between the canal and the river. Instead, goods were to be transferred from the canal to the river on a series of short inclined planes and also via a great warehouse, built straddling the canal and the river bank.

The Hay Inclined Plane still survives – partly because its maintenance was a condition of the agreement to built the London and North Western Railway line through the area in 1857 – and thus it operated until 1894. During restoration, the surface was built up with 3000 tons of new ash. Most of the rails have been replaced to match the L&NWR B190 track put in during the final phase, although wrought-iron bridge rails survive at the top. However, the canal basins at the top and bottom survive intact, as do the remains of the winding-engine house of 1793 and stables at the top. One of the problems at Brierly Hill was that the system was self-winding. At Blists Hill an engine was installed, which archaeological evidence seems to confirm was a Heslop engine.[35]

The canal at Coalport was backfilled early this century, but was partially re-excavated by the Ironbridge Gorge Museum in conjunction with Telford Development Corporation. The exercise revealed the remains of the Watt Meadow stop lock of 1792, and reopened the old water channel linking the top canal to the bottom, but did not expose the extension arm of 1810.

The construction of canals in the Gorge was an expensive and difficult operation, which from field evidence did not bring new settlement or investment on a scale commensurate with that in other areas.

Riverside wharfs

In contrast to the evidence for very heavy investment in other forms of transport, what is most notable about the River Severn, on the south bank at least, is the lack of wharf installations (fig. 4.9). Reynolds' riverside development at Coalport, and the Coalbrookdale Company's Loadcroft and Meadow Wharfs stand out as the few documented loading places, but there are few other visible remains. And yet the number of boats using the river, bargemen, repair yards and boat builders was historically very high.

In part this may be a problem of archaeological survival. The archaeology of the riverside on the south bank and parts of the north is complicated by the dumping of vast quantities of industrial material, including roof-tiles, mining waste and iron-

4.9 'Ironbridge': a heavily laden cart hauls logs up Tontine Hill, while trows may be seen moored at the riverside. A stage coach can just be seen on the bridge (IGMT1978.73)

working waste. Some of this was deliberate, to counteract the problem of slipping. The make-up of the old Salthouses road, for example, comprised several metres of compacted tile waste and furnace slag. This was particularly noticeable in excavations at Ladywood where several metres of cinders and ash covered an early-nineteenth-century settlement, and on the south bank opposite mining waste was tipped across the whole area. The river itself is actively cutting into parts of the banks, and archaeological material can be seen eroding out at Salthouses, at Jackfield and at Ladywood.[36] Finally, the construction of the Severn Valley Railway, with its combination of cutting and levelling along the south bank, seriously affected the survival of riverside archaeology.

River levels fluctuate by several metres each year, and winter flooding is common. There were major floods in 1770 and 1795, alternating with periods of very low water; often the river was not navigable for three or four months of the year. Navigability was declining in the late eighteenth century, blamed by Thomas Telford on the draining of riverside fields. And although a Navigation Bill was proposed in 1786, it was blocked by local interests. There are few weirs, and little control over levels. Such fluctuations may explain why a lock was never constructed between the river and the Coalport Canal, and why the river was rarely used directly as a source of water-power.

The combination of poor survival due to dumping, and the variable nature of the river itself explains why many of the arrangements for unloading cargoes were very casual, involving little more than a bit of land at the end of a path.

Nineteenth-century photographs commonly show boats tied to the riverside, with little more than a plank between the boat and the bank. The Tithe Map shows each of the brickworks of Jackfield as having a path and a small enclosure on the riverside, but most of these were cut by the Severn Valley Railway; today there is little evidence for built wharfs at these sites. Each of the early coal-mines also seems to have been linked with an area for loading on the riverside, probably little more than a track or incline, terminating in an enclosure. Little or nothing can be seen of Clifford's 'four wharf places' of 1608 in the Ladywood area, or of the wharfs at the ends of the main tramways down from Benthall, Buildwas, the Tuckies and Tarbatch Dingle.

Nevertheless, it is possible to identify some more formal wharfs on the south bank. A 2-metre-high stone wall remains at Lloyd Head – the terminus for a group of tramways through Jackfield which continued to operate into the nineteenth century. There is also archaeological evidence for wharfs at the Calcutts. The bank is now being eroded by the river, but traces of wooden piles in the river and sandstone walls, retaining dumped layers of ash and slag, suggest a large installation, creating a level platform on the riverside beneath the steep bank. The site was the terminus of the Jackfield rails, and was used for loading coal in 1748. The Calcutts Ironworks loaded boats there and in 1840 the ironmaster James Foster and the brickmaker Hezekiah Hartshorne used the area. There are also stone walls at Bower Yard on the south side of the Iron Bridge.

The major wharfs were on the north bank. Little survives of the Meadow Wharf, but at Loadcroft Wharf – a railway terminus from the late seventeenth century onwards – the elaborate Severn Warehouse and a loading ramp (now substantially rebuilt) can be seen. From the Severn Warehouse along the whole of the Wharfage to the Iron Bridge, a retaining wall of sandstone blocks survives (see below), broken by several ramps leading down to the water's edge.

The most important wharf remains are those of William Reynolds' Coalport development, built in the 1790s. Although the proprietors of the Preens Eddy (or Coalport) Bridge were given permission to construct wharfs 60 yards either side of the new bridge in 1780, there was little or no development by 1788.[37] The completion of the canal alongside the river by 1793 involved major construction work along the banks. A massive stone retaining wall was built, extending from the Coalport Bridge at least as far as what is now the Brewery Inn. It comprises large sandstone blocks tied with iron cramps and, at the western end, supported on brick arches. Set into the wall are the foundations for the stone warehouse, built spanning the canal.[38] These walls are rapidly collapsing and in need of urgent conservation. The stonework and the remains of the warehouse are evidence for the level of investment, and the vision of Reynolds, which was perhaps not borne out by the long-term success of the venture.

The archaeology of the riverside is vulnerable and often difficult to investigate. Many of the links between transport routes and the riverside have been lost through erosion, and others have been obscured by later dumping. Yet the use of the river is documented as a major economic activity. It has to be accepted that in this instance the documentary evidence may be a better source for levels of activity than the archaeology.

Bridges

The major communication problem within the Gorge was, of course, crossing the river. If the physical nature of the river made the construction and the survival of wharfs very precarious, it made the bridging of the Gorge even more difficult. Problems caused by flooding and erosion were further exacerbated by the unstable nature of the river banks caused by natural slippage and later mining. There were few natural crossing-points for the river which were not affected either by unstable banks or by uncomfortable approach routes.[39]

Physical restrictions were one reason for the lack of permanent river crossing before 1779, but the second, and most important, was demand. The seventeenth-century coal industry had been export-based, and the tramway system was orientated towards the river. However, the need to cross the river increased dramatically with the integration of the large ironworking concerns and the change in focus of the coal industry from export to local supply. Many ironmasters held mineral leases on the opposite bank, or obtained bricks and supplies from distant works. The railway network and the road system were growing in response, and a crossing-point was badly needed.

The location of the Iron Bridge has to be seen in the context of the existing landscape. Like the later Coalport Memorial Bridge, the Buildwas Bridge and the Free Bridge, it was built on the site of a ferry. In other words, the site was already an established river crossing-point which had generated riverside settlement and road access to the river bank well before the bridge was built.[40]

Once opened, the new bridge had a major effect on local patterns of communication by bringing traffic for the first time through the Gorge itself. A new turnpike road was built through Leighton to Shrewsbury to compete with the Much Wenlock route. At the bottom of Coalbrookdale, the old road was straightened and a new brick bridge built over the stream in 1781.[41] The road along the Wharfage was improved, and a new road built up what later became Church Hill, and in 1828 the struggle to maintain the road on the south bank was abandoned, and a new route built up to Broseley. Coaches running from Shrewsbury to Bridgnorth were diverted through the Gorge to view the new bridge, and local inns profited from new customers. Prior to the construction of the bridge, what is now the town of Ironbridge was little more than a scatter of houses along the Wharfage, and a straggling suburb of Madeley at the top of the hill. A market-place, and a new inn, established soon after the construction of the bridge, became the focus of the new town (see chapter 6).

The Iron Bridge still stands today, despite several attempts to demolish it.[42] The south abutment – originally stone – was reconstructed first in timber, and later in iron, as a result of flood damage and pressure from the banks. The original black ashy surface has been replaced by a modern gravel finish, and the railings, once coated with mineral tar, have been painted. A mirror-image concrete arch under the water strengthens the iron arch and prevents the abutments from moving together.[43]

There is a toll-house on the south side of the bridge, and contemporary with an extension to the building a second single-cell building was constructed to the east. Although undocumented, the structure was part of the gate emplacement, and seems from its position to have related to payment for foot passengers or stabling. The Tontine stables to the north-west have been demolished, although their stone-built foundations remain.

In 1779, the same year that the Iron Bridge was built, a timber bridge was built at Preens Eddy, near Coalport, and completed in 1780.[44] The stone abutments on the north and south sides today relate to this early bridge. It was damaged in 1791, and repaired with iron ribs after floods in 1799. The bridge was rebuilt in iron in 1818, and is one of the earliest iron bridges still in commission. The bridge was a speculative venture as part of the general development of Coalport, and unlike most of the other bridges did not relate to a pre-existing river crossing. There were plans by the proprietors to build wharfs up and downstream, which were only partly fulfilled, and the road link to Broseley on the south side was not completed until 1797.

No other road-bridge was built across the Gorge until 1909 when increasing demands for a toll-free crossing and a fortuitous legacy resulted in the construction of the ferro-concrete Haynes Memorial (or Free) Bridge, once again on the site of an old ferry (Adams' Ferry) (fig. 4.10). The bridge is of significance as a fine example of Mouchel-Hennebique ferro-concrete construction and, whilst not the earliest in Britain, is one of the few such structures to be listed.[45]

4.10 The Free Bridge, Jackfield, soon after it was constructed (IGMT1980.1645)

Since then three further river crossings have been built in the Gorge. The 1922 footbridge at Coalport was built as a memorial to the dead of the First World War on the site of the Coalport (or Werps) Ferry. The bridge linked the working communities of Jackfield with Coalport – a major source of employment. Two further bridges are associated with the Ironbridge Power Station: the Pratt truss steel bridge of 1932 and a concrete road-bridge linked to Ironbridge B station, which opened in 1969.

With the exception of the power-station bridges and the Coalport Bridge, each bridge that has been built across the Gorge has been located on an established crossing-point at a point of clear demand. However, once a bridge has been built, there is almost always pressure for improvements in the associated road network. At Coalport, Ironbridge, the Free Bridge and most recently the Buildwas Bridge, road improvements followed bridge construction, and the bridges continue to exert a powerful influence on road policy. The location of the very earliest ferry crossings continues to influence the form and functioning of the transport network today.

Turnpike roads

The construction of roads in the Gorge cannot be divorced from the pattern of river crossings. Once built, bridges exerted a powerful influence on the pace and scale of road building and improvement. Nevertheless, roadmakers faced the eternal difficulty of getting from the riverside up and out of the Gorge, and the solutions they adopted were to have a significant effect upon the subsequent development of the landscape.

Plymley was not impressed with local roads. 'The roads in this country, both turnpike and private are generally bad; the private ones . . . are almost impassable.'[46] The roads of the Gorge were notoriously bad, and, as the minutes of the Proprietors of the Iron Bridge show, constantly in need of repair. Even today, the Coalport road moves continually, and the old road through Jackfield has completely collapsed into the river. The problems today are twofold – natural slippage on the steep slopes, and the legacy of past mining activity – and are likely to be the same problems as those that faced earlier road-builders. Thus road-building in the Gorge has always required technical ingenuity.

Despite the development of tramways, and improvements in the canal network, local firms continued to rely upon road transport using packhorses or wagons. The Benthall potteries, for example, are well away from the river on the south bank, and must have been sending material into Wales by road. However, the most graphic evidence for the continuing importance of road transport comes from the distribution of Broseley pipes, many of which must have been transported overland.[47]

The medieval road network largely bypassed the Gorge to the north and south, and there were few cross-routes until the construction of the Iron Bridge in 1779. The earliest routes down into the Gorge seem to have been little more than tracks. Paradise in Coalbrookdale was an old bridle way, and there were tracks down through Madeley Wood and what was to become Ironbridge. On the south side of the river, there was a Horseway down to Adams' Ferry by the seventeenth century,[48] and routes down to the river from Benthall and Broseley Wood. There are also traces of routes further to the east. Such roads were a matter of custom, and received little maintenance. Few have left any trace on the ground unless they were later adopted as turnpike roads or county roads and improved.

The road from Shifnal to Much Wenlock crossed the Gorge from east to west, leading from Buildwas Bridge along the river, up Lincoln Hill towards Madeley. The original route of this road can be seen a little to the north of the current route, and is reflected in the line of early cottages at the bottom of Coalbrookdale. It was turnpiked in 1764 but the diversion at the bottom of Coalbrookdale dates to the 1780s, when a new bridge and a turnpike gate were built.[49] The road was particularly steep up Lincoln Hill, but because it ceased to be a turnpike it now retains its eighteenth-century character; it is narrower than a modern road, with a sharp junction onto the Wharfage, and an almost impassable climb.

In the latter half of the eighteenth century – partly as a result of the road-building powers given to bridge proprietors – a group of new turnpikes were constructed through the Gorge. Some of these were based on existing lanes; others were new, purpose-built roads. Map evidence makes it possible to draw a clear distinction between the two.

On the south bank, the proprietors were able to use the route up Bridge Bank, but on the north bank a new road was constructed.[50] This became Church Hill, and the map shows the road cutting through earlier rectangular plots to create oddly shaped ones.

Madeley Hill (1810) and the road up to Broseley (1826) were both built in response to continuing problems with the steep routes up the valley sides from the Iron Bridge. Neither followed an earlier route for more than a short distance. Both are straight and cut across the hillside, providing a gentler slope than their precursors. Both cut through existing field boundaries, and the stratigraphic relationship with earlier buildings and roads is awkward. The Broseley road sweeps up to the east of Broseley to meet the Bridgnorth road, the Madeley road to meet the turnpike road at Hilltop.

As with late-eighteenth-century tramway routes, the scale of engineering of these two roads stands in contrast to earlier roads. Madeley Hill is supported on brick arches and the Broseley road cuts diagonally across a hillside on a large earth embankment, both of which required considerable ingenuity. We know nothing of the surfacing, although the earlier roads up from the Iron Bridge (and indeed the bridge itself) were surfaced with ashes and slag from the Benthall Ironworks.[51]

It was rare that roads and tramways shared the same route.[52] The right to lay rails was usually included in a mining lease, with the rails on private land; and, at least locally, it was not until the construction of the mainline railways that parliamentary Acts were passed. Edward Harries had given up Benthall Rail when the road was turnpiked (although in the 1880s rails that still crossed the road created continual problems by overturning carriages). Otherwise, most railways followed totally separate routes. Occasionally, old railways became public footpaths, or even (as in the present course of Coach Road) public roads.

The situation in Coalbrookdale was highly complex. The earliest mapped road ran up the valley, around the Upper Forge Pool dam, and dropped down into the works opposite the chapel. A parallel tramway route was built through the Dale, but did not share the carriageway. In 1817 the road was turnpiked, the new road following the old road as far as the chapel, and then being diverted in front of Carpenter's Row, effectively cutting through the gardens. Even in 1902, most of the tramway route through the Dale was grade separated from the road. The distribution of early settlement suggests that the road through the Dale was largely used by the works; early cottages are to be found along the old course of the coach road to the north-west of the Dale, on Paradise and Church Roads.

Equally complex is the Wharfage in Ironbridge, today a single road made up of several original features – a turnpike at the bottom of Lincoln Hill, a private quay belonging to the Coalbrookdale Company along the riverside, a lane known as Tontine Hill, part of the Ironbridge market-place and finally Madeley Bank. As part of the deal with the Coalbrookdale Company to use the Wharfage, the proprietors agreed to build the wall to widen the road – presumably the large sandstone walls which at present form the riverside.[53]

The pattern of main roads through the Gorge today is largely the result of the early-nineteenth-century improvements following the construction of the Iron Bridge. Madeley Hill, the Wharfage and the road up to Broseley form the main routes – Bridge Bank and Lincoln Hill remain as awkward and difficult as they were in the eighteenth century. It is still not easy to cross the Gorge from north to south, and to this end a bypass has been built to the west to provide for much of this north–south traffic.

The standard-gauge railways

In the mid-nineteenth century, the Gorge was still crossed by two eighteenth-century bridges. The Midlands were accessible from the Gorge by road, and by one branch of the canal operated via a cumbersome inclined plan. Otherwise, the river remained the main method of transport out of the Gorge. The nearest centres of population, and thus markets for industrial products, were Shrewsbury and Birmingham, and transport costs were beginning to dominate the costs of production.

The linking of the Gorge to two national railway networks – the London and North Western Railway and the Great Western Railway – ought to have solved these difficulties and brought new opportunities to the area. In 1862 the Severn Valley Railway was opened from Hartlebury to Shrewsbury along the south bank, and in 1864 the link to this line from Lightmoor through Coalbrookdale and across the river was opened. An L&NWR line was completed along the old canal and down the Washbrook Valley to terminate at Coalport in 1860.[54]

The construction of the railway had a devastating effect upon the community of Jackfield. The railway cut through a densely populated area, and at least 50 houses were demolished. Others lost gardens or were separated from their plots of land. Elsewhere in the Gorge the impact was minimal, and the routes managed to avoid major settlements. Nevertheless, the construction of the railway required a major engineering input in order to overcome the physical problems of the Gorge. Viaducts carried the Great Western Railway route down Lightmoor Dingle, across the Upper

Furnace Pool and along the side of the Dale. The Albert Edward Bridge, designed by John Fowler, was built to carry the line across the river to link with the Severn Valley Railway.[55] Along the south bank, a large viaduct clings to Benthall Edge. Subsidence was a continuing problem at Jackfield and further south, often solved by the dumping of industrial waste.[56]

Certainly local industries took advantage of the new railways. Craven Dunnill's new factory was built beside the railway line in 1874, and Maws moved down to Jackfield soon after. Several of the brickworks built railway sidings which can still be seen, and the old wharfs were closed down and dumped upon. Most of the brickworks advertised delivery by rail or by river. To the north, the Coalbrookdale Company built a railway siding, and the Madeley Wood Company at Blists Hill built a wharf to link their own tramway system with the main line. The Coalport China Works now had railway access, as did the limestone workings on Benthall Edge.

The railway would be expected to have had an immediate impact on the river trade, and on the local tramway networks, and yet it was almost 50 years before either totally disappeared. New tramways continued to be built around sites such as Blists Hill and Jackfield throughout the second half of the nineteenth century, and it was only in the 1920s that these networks seemed to contract. The plateway siding at Blists Hill, for example, was only replaced by a mainline siding in the works after 1902. Equally gradual was the impact on the riverside. Huge quantities of waste were dumped upon the wharfs at Jackfield, and the old wharf at Loadcroft became an engineering works in 1901. The railway company was obliged to maintain the Hay Inclined Plane and the operations of the canal, but these barely survived into the twentieth century. Brickworks continued to advertise river transport, and the last trow – the *William* – travelled down-river in 1875.

If the railway brought some prosperity to the brick and tileworks, it did not result in long-term industrial growth or local prosperity. The population of Benthall continued to decrease, the houses demolished in Jackfield were not replaced. No new industries opened, although there was a brief rejuvenation of limestone working at Benthall in the 1920s. The railways failed to have any major impact on the pattern of closure of ironworks, including Blists Hill in 1912.

The Coalport station closed for passengers in 1952 and freight in 1960, and the Coalbrookdale line remains open only for coal traffic to the power station. The Severn Valley Railway station on the south bank closed in 1963 and the line in 1970. Only two of the four stations which once served the Gorge survive, and both have been adaptively reused – one as a wood workshop, and the other as a private house.

Ironically, the solution to the long-standing transportation problems of the Gorge proved of no lasting value. With the construction of railways, access to the Gorge was now relatively easy, but the base of population, skills and demand no longer existed, and the raw materials were no longer economic to exploit. The heroic investment made to overcome the problems of transportation in the eighteenth century had brought short-term benefits, but the continuing investment needed to maintain these systems proved too expensive in the early nineteenth century. By the time the standard-gauge railways were built, the stagnation of industry and population was established and the railway had little impact.

Thus, the pattern of industry cannot be divorced from the network of roads, railways, canals and river transport which supplied it, and these cannot in turn be separated from the physical difficulties of crossing the Gorge. This communication network was part of the human landscape imposed upon the Gorge, and inherited from generation to generation. It continues to exert a powerful influence on the Gorge today. It is possible to argue, therefore, that the roads, bridges and footpaths of today have their origins in the demands of the seventeenth-century coal industry.

Equally, the inheritance of water-power from – in some cases – the medieval period had a strong influence on the eighteenth century, and meant that water-power continued to be important until well into the twentieth century. Yet the physical geography which made water-power possible hindered transport links. As ever, it is important to set industry not just in its contemporary landscape, but in the context of the landscape it inherited from the past.

5

INDUSTRIAL BUILDINGS IN THE LANDSCAPE

Introduction

This chapter is about the principles by which one can use the archaeology of industrial buildings as a source for understanding the process of industrialisation. Industrial buildings and their works are prime evidence, not only for the technologies of different industries but also for the way that firms developed, the cycle of decision-making and the pattern of industrialisation. Industrial buildings have tended to be regarded as a subcategory of architectural history, with particular attention being given to the exceptional, the most ornate, the most unusual, the earliest or the largest. Architectural historians of industrial buildings are more often concerned with the dichotomy of function and ornament, whereas the industrial archaeologist places the main emphasis on process and the development of new building materials. These approaches only come together where a building illustrates through its materials an 'industrial' development.[1] The availability and use of materials, both traditional and modern, will be seen not as an end in itself but as one of the many factors in the evolution of industrial building form.

Much of the history of industrial buildings has been seen in terms of the textile mill – the archetypal factory. Yet there are no such mills in the Gorge, and few of the typical industrial works to be found here could be described as factories in the traditional sense, nor did they even fall under factory legislation until the 1860s.[2] The Gorge therefore presents a rather different view of industrial buildings.

A works evolves not as a series of single buildings, but as a complex of different but related features, typically centred around an existing resource, such as a mine or tramway, wharf or pool. The changing nature of the industrial structure offers a basis for understanding the evolution of industrial complexes. This, in turn, may lead to insights into the directly practical aspects of running a works, in terms of basic process, finishing, storage and more subtle factors such as advertising, labour control and the expression of pride and confidence.[3] At the root of the analysis of these patterns is the issue of whether the concept of the factory is a useful one; or whether it might not be more useful to see the process in more flexible terms, as the development of works made up of distinctly industrial buildings.

Traditionally the evolution of industrial buildings and sites has been discussed in terms of the emergence of the factory (typically the textile mill) – a centralised unit for mass-production, power, control of labour. This approach very strongly emphasises the central structure and its appendages. In the Gorge and many other regions where there is no great tradition of textile-working, it may be that the idea of the industrial works is a more apt framework for discussion. The works may be defined as a group of industrial buildings or structures, sharing a common purpose and enclosed by a boundary wall or by an area leased or owned by a particular business.[4]

The structures which make up an industrial works can be fruitfully divided between those which are specific to a particular process (for example, the blast furnace, kiln or engine house) and buildings which are much more general in their conception, primarily concerned with protection from the weather, and the provision of light and ventilation. There has been a tendency to elevate the importance of the former as the focus of investment; they have the advantage of being readily identifiable remains whose retention can be justified more easily than an apparently anonymous shed.[5] Yet the surviving evidence of the highly specific structures of the industries within the Ironbridge Gorge indicates that they are not as central to the working of industry as has been assumed. Unlike structures such as windmills it is not justifiable to study them in isolation. Although a blast furnace may be the focus of an eighteenth-century ironworks, ceramic kilns are short-lived, and their remains may be scattered in different areas within a works. The question may therefore be asked whether, during the eighteenth century, there emerges a distinctive ubiquitous, industrial building type, apart from the specific kilns or furnaces, and, if so, what implications it has for the pattern of industrialisation.

Here in the Gorge, the industrial building type marks not a revolution but an evolution from a strong vernacular tradition, largely rooted in domestic and agricultural architectural practice.[6] It may be more appropriate to trace the emergence of a vernacular tradition in industrial buildings – perhaps a works or 'factory vernacular'.[7]

Early industrial structures

It has been shown in chapter 2 that local industry must be seen in the context of events in the sixteenth and seventeenth century. Chapter 8 details the pattern of local vernacular housing, whereby a tradition of timber and stone construction was suddenly, and very thoroughly, replaced by the almost universal use of brick in the small workers' cottages of the early eighteenth century. Far fewer industrial buildings remain from this period, but traces of a similar transition can nevertheless be seen.

One building which may well be associated with seventeenth-century ironworking in Coalbrookdale is Rose Cottages, a two-storeyed timber-framed group of cottages, with a dated stone porch of 1636 added to it. An agricultural origin has been suggested by the fact that it is wider than most similar examples of domestic buildings,[8] but it

5.1 The Malthouse, Coalbrookdale, the possible site of Basil Brooke's steelhouse (IGMT1982.256)

sits at the base of one of the Coalbrookdale pools, in a good position to utilise water-power.

A second building of possible industrial origin in the Dale was the 'malthouse, formerly a steelhouse', located at the south end of the Upper Forge (fig. 5.1). Demolished in the 1960s, photographs and plans show that it too was timber-framed, and set adjacent to a source of water-power. This was certainly part of the seventeenth-century Great Forge complex, and either this or Rose Cottages may have originally been the 'Calbroke Smithy' of 1536.[9]

A third potential industrial structure was 15/15a Holly Road, Little Dawley, just to the north of Coalbrookdale; adjoining a small, late-sixteenth-century cruck-framed building was a seventeenth-century box-framed structure.[10] In the yard outside, remains of an ironworking bloomery with no related buildings were found in association with medieval pottery.

Evidence for early industrial buildings must therefore be sought further within the tradition of large timber-framed buildings, which may in turn relate to techniques of agricultural barn building as a solution to covering large areas. It is also possible that there was an early industrial building tradition using stone. Abraham Darby I employed 'masons' and 'tilers' in the early days at Coalbrookdale which might imply the use of stone in building. Local seventeenth-century domestic structures made use of coal-measure sandstone and there are remains of seventeenth-century stone barns in the area (for example, that at Hay Farm).[11]

But it was the use of brick and tile which provided the wherewithal to construct large, reasonably fireproof buildings for industrial purposes in the eighteenth century. Brick was robust, relatively cheap and could be built to a greater height with less mass than was possible with stone. It was ideal for large simple structures and allowed for greater flexibility in openings. The popularity of brick as a building material was not confined to the brick cottages, built in large numbers for an increasingly industrialised population from the early eighteenth century. In less domestic contexts it can be seen in the tradition of large-scale brick and tile barns of the mid-eighteenth century, and on a smaller scale in the brick-built sheds to the rear of a domestic property, serving as a brewhouse or, more rarely, blacksmith's shop. Thus the simple, gabled brick structure was well established as a local vernacular tradition by the mid-eighteenth century.

The development of the ironworking shed

The development of an industrial vernacular can be seen most clearly in the ironworks of the mid-eighteenth century. At this date, there is clear evidence for a series of large, gabled brick buildings, with tiled roofs supported by substantial timber trusses.

While some industries could readily operate in the open-air, the bulk of the stages in ironworking needed to take place under cover. Processes such as hammering, casting and reheating necessitated that any buildings should embrace a relatively large area, should be well lit and ventilated, and should be relatively fireproof. The buildings should also be strong enough to withstand the vibration of hammering, and intense temperature changes. There was no need for the centralised power supply of a textile mill, the well-lit bench space of a workshop or weaver's cottage.

The best examples of the eighteenth-century ironworking shed were the buildings at Newdale, about 6 km to the north of the Gorge, where a pair of well-preserved buildings dated to 1759–60 stood.[12] They were built by the Coalbrookdale Company as a model ironworks, on a site which had no readily available water-power. Two double-gabled brick buildings remained, one in use as a farmhouse, the other as a barn. Archaeological investigation of the buildings prior to their demolition showed that, in their original form, the buildings were open to the roof. A massive iron valley beam separated the two bays of the buildings, and this was supported on a pair of cast-iron columns in the centre of the room, creating a large open space. The buildings were lit by a series of small rectangular windows on the north-east and west sides, and by larger rectangular openings in the east gable wall. Kingpost trusses supported a tiled roof. Outside the original openings to the buildings were the remains of two air furnace bases, filled with slag and iron scrap. There was archaeological evidence for another industrial building to the north-west, with open arcaded sides, and central hearths with a pair of pits which seem to have been used as water baths to cool iron.

The two Newdale buildings provided rare evidence for the appearance of eighteenth-century ironworking buildings rather than furnaces (fig. 5.2). They mark the very early use of a hybrid mixture of wood and iron in construction – cast-iron was used to support the building and in the valley gutter instead of the more usual timber. The high roof provided ventilation, and minimised the risk from sparks, while the large

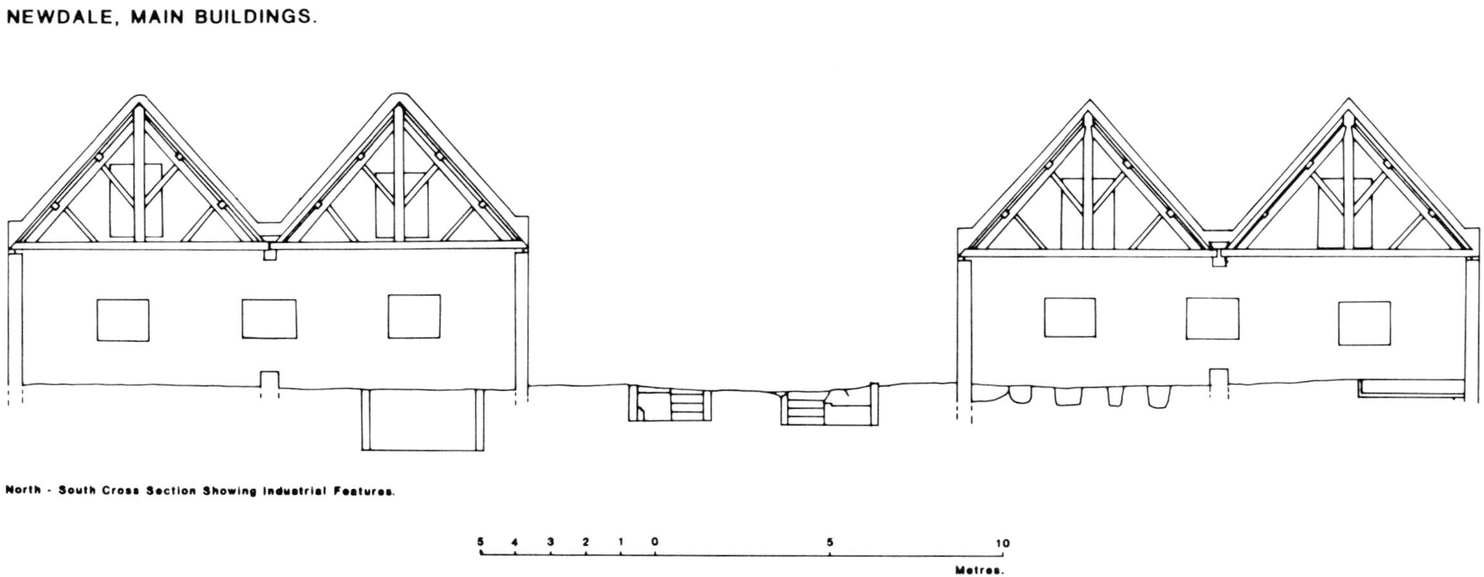

5.2 Section through the pair of ironworking buildings at Newdale showing the underlying archaeology. A casting pit lies within the north building, while a pair of air furnaces may be seen between the buildings (Mark Horton/Nicola Smith)

windows provided ample light. The double gabled form was copied in the roof of the row of back-to-back houses built nearby at the same time. Both sets of buildings were on a similar alignment, suggesting that the site had been conceived as a highly organised community set within a rectangular area cut out of the surrounding fields.

This formula emerged again at the Upper Forge, Coalbrookdale, where there remains a single-gabled, tall brick shed, lit by a circular opening under the south gable. It was built soon after 1753 as a forge, and by 1786 a hammer engine had been installed in the strong brick tower which abuts the north end. Subsequently the forge was extended to the south, and in the nineteenth century a water-wheel was inserted and the roof partly raised as a mill and stable for the Coalbrookdale Company. Later a large opening was cut into the front of the building during its use as a garage and plant hire firm. The present roof is not primary, and an illustration of 1786 shows the original to have had small slatted vents. Once again, multiple uses illustrate the flexibility of this building type.[13]

One unusual feature of the Upper Forge is the use of arches; semicircular relieving arches are built into the fabric of several elevations of the original structure, either to carry it over a former watercourse or to strengthen a building in which cranes were used. In the front façade (now largely rebuilt), there are traces of a primary opening in the form of a much shallower arch, perhaps used to remove large sections or plates of wrought-iron from the building.

On a larger scale, the Soho foundry at Smethwick, built by 1807,[14] uses huge arched openings on the boring mill, with further arches on the foundry, pattern and turner's shops. Elsewhere it is used for abnormally large doorways, such as the Dukes grain warehouse and Castlefields warehouse, Manchester. But it is the Horsehay Ironworks in the Gorge which provides evidence for one of the most splendid arcaded ironworking buildings.

The New Rolling Mills at Horsehay, just to the north of Coalbrookdale, mark the development of the local industrial building type to a much larger scale to contain the new technology for rolling wrought-iron.[15] Built as part of the reorganization of the works in 1806, the building was a triple pile, brick and tile structure, lit by three tiers of iron-framed windows with a primary arcade at ground level which was later enlarged. At one point, Horsehay was capable of rolling the largest boiler-plate in the world; it specialised in rolling sections of different shapes and sizes, producing up to 500 tons per month. The arches would have allowed the easy removal of large boiler plates from the building. The rolling mills were thus a key feature of the site. The mills were steam-powered from the outset, and contained four trains with reheating furnaces. The original engine was replaced in 1849, but must have taken water from the pool for condensing. An adaptable building, it was converted into a machine shop in the 1890s; it was reroofed in 1902 and again in 1838 and finally demolished by 1956.

An illustration of 1858 shows such a building in operation – balls of iron are placed under a hammer, iron plates are fed through paired rollers onto a small cart, whilst bar iron is passed by hand over a pair of rollers to be rolled again.[16] The picture clearly demonstrates the need for space in a rolling mill, solved so well in the structure through the use of multiple gables and iron support. Yet from photographs, the tiered windows and arches give the impression of a very much larger building than the known dimensions suggest – it was just 23 feet to the valley gutters – and may have been a deliberate attempt to increase the impression of size.

The rolling mills at Horsehay were designed by Edmund Darby, then manager of the works, and built to his specifications by Messrs Wetickliss & Sons, a local Ironbridge building firm.[17] This is one of the few industrial buildings whose designer and builder are documented, and it is significant that they are both local. Otherwise, company accounts make reference to the supply of bricks and the work of masons, and we know that the ironworks invested in brickworks for their own use, but there are few hints as to who conceived or built these structures.

The Erecting Shop to the south of the Great Warehouse is a direct successor – albeit a more specialised one – of the eighteenth-century ironworking shed.[18] Built by 1882, this is a single-storey brick shed with hipped roof, and a span much greater than could be achieved through the kingpost truss. Inside, the additional height of the building allowed for the track of a travelling overhead crane to be built into the fabric of the structure, and the use of piered construction lightened the mass of the walls. The interior was lit by large, arched, multipaned windows in the eastern and western walls. Interior photographs *c.*1900–5 show row upon row of pumps, each being assembled by an individual operator, not yet incorporating the linear organisation of mass-production (fig. 5.3).[19]

If the erecting shops demonstrate the growing need for height and space in the ironworking shed, the Severn Foundry of 1901 continues the trend towards enclosing larger and larger areas. Four ranges are placed at right angles to each other, to create a vast open interior supported on iron columns and lit by clerestory windows. Considerable extra width was gained through the use of the queenpost truss – an eighteenth-century feature which became standard on many other industrial buildings such as the Coalbrookdale Long Warehouse[20] – and extra light came through large multipaned windows. The foundry was built to accommodate the production of gas cookers for the London council-house building programme, but the demand was short-lived and it closed during the First World War. It is now used for the manufacture of teddy-bears, showing just how adaptable the industrial shed could be.

The alternative solution to enclosing a large area was the use of multiple gables on the distinctive 'saw-toothed' building at the southern end of the Coalbrookdale site.[21] There a single-storey building of the late nineteenth century has a rather ponderous nine-ridged hipped roof, each ridge with north light. It is supported by iron columns, cast to resemble girders. Interestingly, the building is three-sided, the rear wall being originally open, suggesting that it would have led directly into another covered ironworking area. This adjoining structure has since been replaced by a more recent one.

The multiple-ridged roof was also taken up by the company in the North Lights building, keyed into the rear of the Long Warehouse and built in the early twentieth century. In fact the concept of the north light shed had been developed in textile mills by the 1850s, and was relatively late in the Gorge, underlining the general impression that the pace of innovation set in the mid-eighteenth century had declined

5.3 The interior of the erecting shop, Coalbrookdale (IGMT1981.3020)

by the nineteenth century, when Coalbrookdale no longer played a major role in national ironmaking.

It is apparent from the buildings used by the Glynwed Company today for casting and finishing that the evolution of ironworking buildings has culminated in most production now taking place within open sheds, with high roofing and unbroken floor space. Asbestos, coated profiled steel and plastic cladding on a steel frame were adopted rapidly by the iron industry and represent the ultimate expression of the generalised industrial shed, providing little more than light and cover from the weather. The last big brick buildings were constructed at Horsehay in the early twentieth century, and thereafter brick was used only for offices or specialist structures. It may be that the little-appreciated ironworking sheds of the Ironbridge Gorge represent a precedent for the anonymous inter-war production area, and for the modern steel-framed industrial unit. Clearly it is important that the evolution of an early industrial vernacular in the eighteenth century, and its later development, is explored more fully for other regions and other industries.[22]

It has also become apparent that the development of the single- storey ironworking shed in the eighteenth century was very different from that of the multistorey textile mill. The multistorey ironworking building was the exception; it tended to be associated not with primary manufacturing, but with warehouses, storage or drawing offices or, as at the Soho Manufactory, the upper storeys were used for workers' accommodation.[23] The emphasis on power and machines, and their incorporation into centralised operations has meant that the history of industrial building in the eighteenth century has concentrated on the development of the great textile mills. Yet the buildings of Coalbrookdale, Horsehay, Newdale and no doubt many other areas demonstrate a very different tradition – less pretentious, based more in local practice and concerned with the problems peculiar to ironworking.

Although they share many common features, the warehouses which survive today are in a very different tradition from that of the industrial shed. The Great Warehouse at Coalbrookdale (1838) and the Severn Warehouse on the riverside display a combination of the structural confidence visible in other industrial buildings, with the take-up of features of a more heroic industrial tradition – that of the multistorey warehouse or mill.

When a new letterhead was drawn up for the company in the 1844,[24] it featured the Great Warehouse – a three-storey building of nine bays, its double-pile roof surmounted by an ornate cast-iron clock added in 1843 (fig. 5.4). Iron columns support the massive wooden floors and are used in the lintels, sills and frames of the

5.4 The Coalbrookdale Works in the 1840s. The newly built Great Warehouse can be seen in the middle distance (CBD59.82(22))

doors and windows. Until *c.*1827, the Upper Works had been a jumble of structures, which had as their cores the buildings of 1753 and earlier. In the 1830s, the site was reorganised and began to move into the production of art castings. The Great Warehouse lay in the centre of the site, visible from further down the Dale, and acted as an elegant front to the older buildings.[25] If the model for the industrial shed was akin to a barn or small workshop, the basis for the Great Warehouse was very different;

many elements are local, but the scale of the building and the clock express a very different tradition.[26]

Equally anachronistic in the context of local ironworking buildings was the Severn Warehouse of 1838/9, facing out over the River Severn – then the busiest approach to the Gorge. Built of red and yellow handmade brick, with castellated turrets, mock arrow loops and pointed windows, it is so far divorced from the local tradition of the industrial shed that it may well have been architect-designed. The iron-framed multipaned windows are a local feature, but here they are pointed. Even on a national level the building was unusual, being one of the rare examples of Gothic detailing applied to an industrial warehouse.[27] It is worth noting that both the Severn Warehouse on the river, and the Great Warehouse on the road/tramway network, were concerned with the dispatch of products rather than integral with manufacturing. To this extent they stand outside the tradition of the manufacturing shed, and more within the ambit of the pottery entrance or warehouse.

The changing ceramics complex

In isolation, the ironworking shed demonstrates the way in which the demands of an industry can be related to changes in building techniques and materials through one specific emerging industrial building type. However, relationships between structures may actually tell us more about the general process of industrialisation than individual buildings. In the most simplistic terms, the study of the industrial works can tell us about the way an industry has evolved, the changing rationale of industrial management and the nature of work. Perhaps the most fundamental issue lies in trying to define how far managers were governed by a mentality of 'make do and mend', and at what point they took the bolder decisions to build anew, to rebuild existing works or to relocate elsewhere onto a greenfield site.

The ceramic industries provide the best approach to considering the evolution of industrial works, as they represent widely differing approaches to the same basic process – that of burning and preparing clay. There could not be a greater contrast between the Coalport china factory of the 1790s and a contemporary brick kiln, isolated in a field and serving a local farmer. These very different approaches to the organization of ceramic manufacture can be seen in pottery-making, china manufacture, brick- and tile-making and in the production of decorative tiles.

Pottery manufacture

The remains of one of the earliest pottery-making sites in the Gorge stand at Jackfield today. Now much altered, this row of four small cottages once had a kiln attached to their eastern end, and can be identified as one of the three 'mughouses' shown on an early eighteenth-century estate map (fig. 6.5b). The houses provide rare evidence for pottery-making on a domestic scale, where a works consisted of little more than a house or cottage with associated kiln. Archaeological evidence suggests that common domestic black and yellow earthenwares were made here, probably for local distribution. By this date there was already manufacture on a different scale. To the north of the mughouses lay Morris Thursfield's Potworks – a kiln abuts an L-shaped range, laid out in a rough courtyard form behind a house. This type of semi-domestic manufactory has been identified from the Potteries, where it would have included workshops, a drying shed and perhaps a clay preparation tank.[28] Coal and clay supplies would have come from pits to the north, and lead was obtainable from a local smelter.

The courtyard form common in the Potteries from the late eighteenth century can be seen again at the Benthall Pottery, built in 1772 by John Thursfield.[29] Within what is now an agricultural machinery store, isolated walls of the original pottery still stand. By 1902, the entrance at the front facing out onto the road led into a hollow square, with kilns set within it. Clay was brought directly from a pit to the north-east, to what were most likely clay preparation buildings. Amongst the buildings shown on the plan there would have also been areas for slip preparation, moulding and throwing, drying, glazing and firing. It is unlikely that the basic layout of the works had changed much over the previous century, although the kilns were almost certainly altered and rebuilt. This formal arrangement contrasts with the scattered Haybrook Pottery immediately to the south, erected in 1743, and also with John Miles' pottery at Ladywood, where the round potteryware works, the flat potteryware works and an oven were spread along a hillside and on the road leading up from the Iron Bridge.[30]

The fragmented walls standing at Benthall are all that remain of at least seven (and probably more) other earthenware manufactories established in the area by the late eighteenth century.[31] All were supplying a very uniform product – hollow-wares and press-moulded flatwares which were more standardized than the early Jackfield products. Such pottery was basic, utilitarian ware, used in the kitchen or dairy, and contemporary inventories show that these wares were becoming more common. But the increase in the number of potteries was more than just a question of local demand, and it may be that this increase, and the technical standardisation, represent a form of mass-production in pottery-making.

Morris Thursfield's old works at Jackfield remained in use until well into the nineteenth century.[32] The site was briefly taken over for experiments in porcelain manufacturing prior to the construction of the Coalport works, and was briefly used also for fine creamwares and mochaware. By the mid-nineteenth century, standard domestic earthenwares of no great quality were once more in production in what was by now an almost derelict site. The tenants complained to the works owner, Mr Harries of Benthall, about the state of the factory, but did not have their lease renewed, and although several other prospective potters considered the site it was eventually taken over for the manufacture of encaustic tiles (fig. 5.5b), and finally demolished just after the construction of the Craven Dunnill factory on the site in 1874. By this stage, the buildings had become almost incidental to the manufacturing, and much of the value of the site was its proximity to clay resources.

'Ordinary yellow and other common wares' were also made in the mid-nineteenth century at the now united Benthall and Haybrook works. Both the position of these

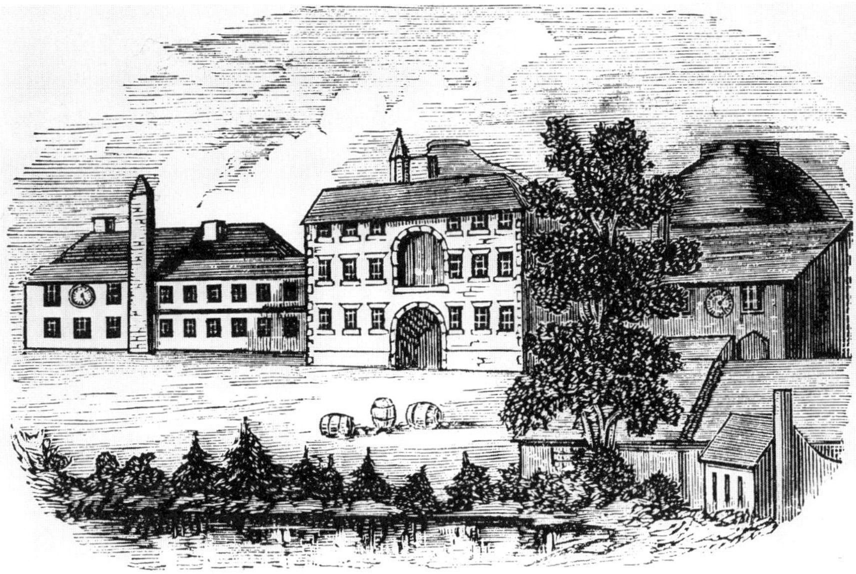

5.5 (*a*) The Caughley China Works, taken down in 1815 (IGMT 1986. 11057)

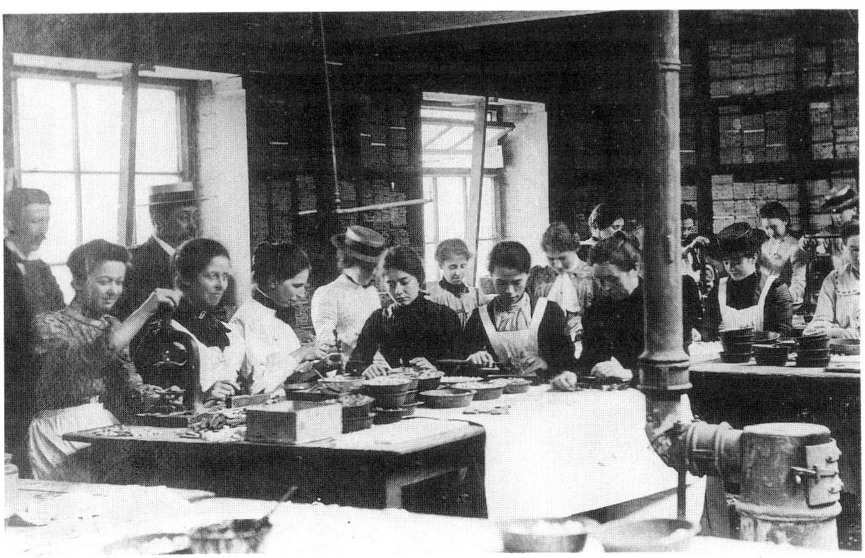

(*b*) Mosaic workers at Craven Dunnill. As at Coalport, the need for well-lit bench space influenced the layout of the buildings (IGMT 1982. 873)

works, well away from the river, and their output suggest a small-scale pottery, relying on fairly basic technology and supplying the varying demands of a regional domestic market – possibly extending into Wales – by road. In 1882, there was a short-lived move into art pottery, perhaps to try to reach a wider market, but the project was never an economic success, and it seems that, while the smaller Haybrook Pottery continued to specialise in such wares, the larger Benthall works moved back towards the production of the old staples – cheap, coarse earthenwares, drainpipes, lamp bases and, later, electrical fittings.

Elsewhere in the country, pottery buildings were as much a matter of image as of function, and it is perhaps not a coincidence that no engravings exist of these earthenware potteries, and barely any of the buildings themselves have survived.[33] What little evidence we have suggests that they were smaller, less elaborate and employed fewer people than the well-known contemporary works, yet nonetheless one can see in this industry many of the trends which were affecting other local industries. Pottery-making had gone through a late-eighteenth-century revolution in production methods through the use of plaster moulds and in some areas steam (which deserves further study), but the building evidence suggests that the industry had never had the confidence to grow or invest in large new works. The trend towards mass-production is visible in the eighteenth century, and, as in other industries, a move towards prestige production in the late nineteenth century was backed by more mundane output. Yet neither conspired to make the Ironbridge Gorge a major centre for pottery manufacture.

China

The large china manufactories at Caughley and Coalport fit much more neatly into a conventional pattern of eighteenth-century industrialisation, with large centralised factories employing many people and supplying a growing national market.

At Caughley on an isolated hilltop about 4 km to the south-east of the Gorge, Thomas Turner of Worcester in partnership with Ambrose Gallimore, in 1772/5 built a complete manufactory, equal in scale and pretension to any at that time in urban Staffordshire.[34] The works were in the form of a hollow square; the main entrance was a large arch in the centre of a three-storey building, with decorative quoins and a hipped roof (fig. 5.5a). Above the main arch is another arched window, presumably for loading raw materials into the upper floors. The buildings to either side, and presumably those to the rear, were notably plainer; on one side was a two-storey range with a hipped roof, multipaned windows and a clock and on the other side was a single-storey building, again with a clock set into the façade. Two very large, fat bottle kilns can just be seen to the rear, and across the road in the front of the site is a pool with several sheds beside it.

Although there was plenty of space for building in the fields at Caughley, the factory was built in an area that was already industrialised. Gallimore had operated an earthenware works there, built originally in 1751,[35] the coal-mines which Turner later leased were well established, and access to wharfs and a warehouse on the riverside was by an existing railway. To the south, sitting on an outcrop of fireclay, was the T-shaped saggar works, with two kilns, which were later incorporated into the works.[36] As there was no power locally, raw materials for the works were ground at the smithies, on Linley Brook.

Caughley was conceived as more than a porcelain works – the design was carried through into Caughley Place, an apparently highly ornate 'chateau' built by 1795 above the works and said to have been designed by one of the French workmen. The only building which remains standing on the site today is equally unusual – it is the 'Roundhouse' to the east, square with blind windows, a decorative plinth of moulded brick and brick quoins. It repeats the motifs of the main works and of Caughley Place, and must have served as a gatehouse to the works, or to the Browne estate on which the whole group was set.[37]

This concern with appearance relates to Josiah Wedgwood's Etruria Works at Burslem,[38] but not to the original Worcester site, completed only three years earlier to manufacture creamware, where the company had taken over Warmstry House and adapted it to their own purposes. There, an ornamental Palladian entrance with clock and bell looks out over the canal, while behind it more mundane buildings form a courtyard. The layout of Caughley reflects the same courtyard plan formalised by Wedgwood and ultimately derived from earlier earthenware potteries. At the Etruria Works, Wedgwood produced creamware, the first British pottery type to be mass-produced and widely sold. Wedgwood's success has been seen to lie very much in the weight he placed on marketing and presentation, and the appearance of his works was part of the image of his product. In the same way, Caughley was built to produce not domestic earthenware but porcelain of high quality for a discerning market. The appearance of the works became part of the image of the product. Skilled workmen were brought from France, and the names of many of the engravers and painters are recorded.

Following Wedgwood's success at Burslem, the manufacture of creamware spread to many parts of England including Coalport, where Walter Bradley operated a creamware manufactory between 1796 and 1800.[39]

Coalport is, however, better known as a china manufactory. Today one range of the original works built by Edward Blakeway and John Rose in 1795 survives (fig. 5.6). It is an imposing three-storeyed brick building of five bays with a central loading door, under a tiled, kingpost and strut roof which was originally hipped. Each floor is lit by windows with single ring-cambered heads and cast-iron frames, but the cast-iron columns and beams inside, and the first-floor windows were added later. The building butted onto the front of another range, presumably of similar appearance, with two kilns, one at either end. It was built by Samuel and William Smith of Madeley, local builders, to the partners' design.[40] In contrast to the ironworking shed, with its image of strength, shelter and ventilation, this is very much a workshop on a human scale. The aim was to light areas where operatives could sit at work-benches in reasonable comfort and warmth. There was no need for the central power transmission of a textile mill, but still the multistorey, centralised, well-lit building prevailed.

5.6 The Coalport China Works early in the nineteenth century, after a painting by Muss (IGMT1986.14088)

In contrast to what was to become a standard pottery layout, the Coalport works were long and narrow in plan, stretched out between the canal and the road. Blakeway and Rose were taking advantage of William Reynolds' attempt to create a transport terminus linking the River Severn with the canal system, and with it an industrial estate.[41]

There were other factors which contributed to the haphazard layout of Coalport. By 1800 a second competing porcelain works had been built across the canal, but the enterprise was short-lived, and by 1814 the site (along with any remains of Bradley's works) was incorporated within the existing works. The form of these works can be seen in an engraving – the original buildings have been extended in a line to the east and west with a jumble of buildings, lean-tos and kilns. Across the canal on the river bank are the two kilns, main building and engine house of the second works. Viewed from the canal and riverside, at least, the works have lost any of the formal pretension they may have originally had, and have become an accumulation of buildings on a restricted site.

One other early building survives to the east of the present site. Originally built as six cottages for workmen by 1797, the building was converted to a warehouse by 1814 (perhaps with the addition of an extra range). This conversion seems to have been very thorough – the classical frontage displays wide sash windows, and pedimented door-cases with fanlights, details which are totally incompatible with the standard artisan's dwelling of the late eighteenth century.[42] The implication is that the roadside frontage was perhaps more important than the riverside, although there is no evidence for an elaborate entrance.

When, in 1902, the manager at the time, Charles Bruff, decided to rebuild the increasingly run-down works, he did so without major alterations to the form of the site, an action which may reflect how few changes had taken place in the manufacture of china in the interim. John Rose's buildings were demolished and rebuilt on the old footings, using iron-framed windows, but with the addition of a bellcote building and showroom facing out onto the road. New machinery was installed, new kilns were built, and the present entrance buildings constructed, yet, as a visitor noted, the 'ancient and the modern mingle in the most harmonious fashion', and old cottages, workshops and kilns survived from the early days.

Very little is known of the way in which processes of pottery production related to buildings at Coalport and Caughley. Yet to some extent these were minor considerations. In china manufacture, the skill and expertise outweighed more practical factors such as the efficiency of the works. Much of Coalport's success was due to the ability of its managers to take over and absorb the materials, skills and workmen of other factories – Caughley, Swansea and Nantgarw were closed and the moulds and some workmen taken to Coalport. Individual artisans were valued, and their names or marks not submerged by the general company image. The firm relied on continuous innovation in designs, colours and forms, yet there were few innovations in their technical processes – all processes were done by hand. Coal-fired up-draught kilns were used, and although steam was applied to blunging and mixing clay, moulding and pressing remained hand-operated.

By the mid-nineteenth century, the buildings of Coalport were neither exciting nor very salubrious, and it is unlikely that the works were very efficient. Nevertheless, as with other potteries, image remained important. Whilst others built grand entrances, Coalport used its location as part of its image – prints of the works were distributed in the same way as Perry had produced engravings of Coalbrookdale, pottery often depicted the Ironbridge or local scenes and the factory itself was at times referred to as Coalbrookdale. Coalport was calling on the traditions of industrial manufacture rather than its buildings to justify itself.

Building products: bricks and roof-tiles

The same pattern of a series of heavily capitalised, specialist operations serving a regional or national market, growing out of a broad-based, local industry can be seen in the brick- and tile-works of the nineteenth century (fig. 6.12). Large purpose-built roof-tile works, with heavy investment in machinery, achieved a spectacular but relatively short-lived success in comparison to more flexible operations on a smaller scale.

Traditionally, brick- and tile-making had been a low-capital industry, with little investment in buildings or plant and little consideration of prestige. Illustrations show a brickworks of 1800 consisting of little more than a few thatched shelters, a straw-covered drying hack and a clamp or kiln of a temporary nature.[43] Work was seasonal, employment casual and the returns not high enough to justify the extensive buildings of, for example, a contemporary chinaworks. The industry was slow to adopt innovations such as coal-fired kilns or mechanical clay preparation, techniques which had long been used in the pottery industry. Brick- and tile-works were not included in the Factory Acts until 1867, and then only after a Commissioners' report on the horrific working conditions.

There is plenty of evidence in the Gorge for the type of small brick- and tile-works common in any settled area of Britain where there was suitable clay. Amongst the coal-mines of Madeley Wood there were isolated kilns, where brick was rapidly replacing wood for props, tunnels and engine houses. Tucked between the houses of Broseley Wood or Ironbridge there were kilns presumably serving a local market and often operated by a local builder. There were kilns in fields as late as 1847, and plenty of field names suggesting earlier brickmakers scattered about the area. The manufacture of tiles (or bricks) 'on the premises for use on the premises only' was probably more common than is documented.[44]

Although this small-scale tradition persisted, it is clear that by the mid-eighteenth century brick- and tile-making was established in a more organised fashion as part of the integration of ironworking. There were brickworks at Horsehay in 1755, and at the same time the Dale Company was providing bricks to sites in Wales;[45] William Reynolds established a brickworks as part of the development at Coalport, and Charles Guest of the Broseley furnaces made bricks at Benthall. We do not know what these brickworks looked like, but it is clear that the demand for bricks made it economic and perhaps necessary for ironworks to produce their own bricks on an industrial scale.

Map evidence shows the striking concentration of brick- and tile-making in Jackfield by 1840: Ladywood, Coalford, Hollygrove and Broseley works were very similar, large works, with a heavy investment in buildings. Each had several sheds, used for clay preparation and for moulding, and presumably for drying.[46] Rectangular and circular structures, located on the downhill side of the works, were clearly kilns and drying stoves. Each operation was located on a tramway or road leading to the river, except two works which were close to Broseley on the Bridgnorth road.

There is little documentary evidence to explain the conundrum of these works – their size, their location and the scale of operation in an industry notorious for its undercapitalised, temporary operations. They were clearly unusual, but none of the works survives.[47] The Broseley Tithe Map describes them as 'brick-kilns', but they must also have manufactured roof-tiles.[48] Neither brick nor roof-tile manufacture was mechanised at that date, but roof-tiles required more specialised processes and buildings – different clay, a different pug mill, better drying facilities, and more controllable kilns. Well before the construction of the Severn Valley Railway with its increased access to markets, there were large thriving purpose-built works here, in an area away from the nearest residential areas but close to the River Severn. The evidence of the buildings seems to indicate the large-scale manufacture of roof-tiles for a regional or possibly national market in Jackfield by 1840.

Between 1840 and 1889 there was an extraordinary boom in roof-tile manufacture fuelled by the construction of the railway, increasing mechanisation and an inordinate demand for durable roofing tiles. In the 1870s, the plastic process was mechanised, and in the 1890s the semi-plastic process was introduced involving drier clay, more machinery and fewer buildings.[49] At the end of the century, there was a sudden and catastrophic decline in the industry.

The best-preserved brick- and tile-making works stands at Blists Hill, on the east side of the canal.[50] Little survives of the three small sheds of 1847, but by 1883 the buildings which stand today were largely complete. The works were set up by the Madeley Wood Company, as a subsidiary activity to their ironworks on the site. Sold after the closure of the ironworks in 1912, they remained in operation until the 1950s.

The Madeley Wood works was conceived as a complete working unit for the plastic process of brick- and tile-making of the 1870s, comprising steam-driven clay preparation machinery, heated drying sheds and down-draught kilns, which were laid out as far as possible to facilitate the movement of clay through the works by gravity. Clay was brought from the mine by plateway to weathering heaps, then hauled into the upper floor of the clay-preparation room, where it dropped through crushing, grinding and tempering processes to the ground floor. There a pug mill, wire-cutting machine and a bat machine for tiles were probably housed. The engine house abutting the clay-preparation room to the east powered the crushing machinery and haulage.

From the clay-preparation room, bricks or tile bats would have been taken to the pressing shed – a large open-sided building, with five furnaces at the north-east end and underfloor ducting. Tile bats would have been dried, pressed and left to dry again. From there the tiles would be moved to the Drying Stove to the east, a building apparently unique to the area, comprising two vaulted bays, separated by brick pillars, the whole heated by underfloor ducts. After 14 days, the tiles were 'white dry' and conveyed to kilns, where they were burnt along with common bricks. Kilns were the most expensive and difficult buildings in a brick- and tile-works, but would last about 12 years if well constructed. It was more economical – and more common – to build them in pairs, sharing a chimney,[51] although there were three original kilns at Blists Hill, all of which were down-draught intermittent kilns with six to eight fireholes protected in typical local fashion by an arch.[52]

A second clay-preparation room (perhaps for the semi-plastic process) was added to the works in 1902, along with offices and another boiler, but as there was little increase in firing or drying capacity, the brick and tile may have been finished across the canal at the newly completed western works. Less complete today, these were purpose-built as a tileworks with an estimated output of 7 million tiles per year.[53] The drying shed is a large rectangular structure with four parallel gables decorated with terracotta finials. Inside, brick pillars support the roof. There was a range of stoke-holes along the front of the building, and heat was carried beneath the floor in ducts. Two brick-vaulted drying kilns abut the west side of the building, both used for secondary drying. The buildings housing the bat machines, and the kilns to the south have been completely demolished, but a post-1902 shed remains to the south.

Other standing evidence for brickworks is sparse,[54] but it is clear that a local style of works had emerged by the late nineteenth century (and possibly earlier), developed to cope with specialist tile manufacture. Down-draught kilns protected by arches, and vaulted drying stoves were local varieties; clay-preparation buildings and pressing stoves were more common features of the Midlands.

Yet this specialisation had its drawbacks. By 1889 the industry was under pressure from slate and from other forms of roofing material, and the quality of the product was deteriorating (ironically because the semi-plastic process may have been unsuitable for local clays). The level of investment in buildings may have contributed to the problem – in 1894 an observer noted that there was 'no lack of capital in the erection of tile-making plants in the Broseley district, some of which would astonish a few who believe in cutting this item down to the lowest possible extremity', but only a few years later there were problems in diversifying into brick, as the works were 'laid out especially for the manufacture of roofing tiles'.[55] Many works had filled large export orders, but with greater competition transport costs became a factor which weighed heavily against the local product, and even a brief revival of hand-made tile manufacture associated with the Arts and Crafts movement failed to revive the industry, most of which closed by the time of the First World War.

Encaustic tiles

The encaustic tile manufactories of Craven Dunnill and Maws were built in response to the extraordinary vogue for tiles as a decorative medium, fuelled initially by the mid-Victorian boom in church building using reproduction medieval tiles, and later by the success of the industry itself, which promoted new techniques and designs through royal and architectural patronage.[56] These manufacturing complexes represent a uniquely documented attempt to combine the circulation of clay and goods around the works in an orderly fashion whilst retaining the elegance and appearance of a pot bank. By the time these specially designed factories were built, encaustic tile manufacture had become a highly complex works procedure, with as many processes as a pottery, if not more.

In the early days of encaustic tile manufacture, tiles were made in different premises. When Herbert Minton adopted the earliest patent for the mechanical production of encaustic tiles in 1830, he 'commenced the manufacture in a single room . . . at the earthenware works, and only three men were at first employed'.[57] Within about five years, there was a short-lived attempt to make encaustic tiles at Jackfield, this time not in a pottery, but in Exley's roof-tile plant.[58] The same tradition of working in existing premises can be seen at George and Arthur Maw's early works at Worcester, where the brothers had bought moulds and equipment from Fleming St John, who in turn was using the old Worcester Porcelain works to make tiles.[59] Later on, the business from which the Craven Dunnill Company emerged operated in an old and derelict pottery.

When the Maw brothers came to the Gorge in 1852, they opened a tileworks on the site of the old Benthall Ironworks. The site chosen was in the Benthall Valley, a narrow, steep site, hemmed about by pit heaps and by waste from the old ironworks. They may have used the buildings of the ironworks at first, but by the time they left the site in 1883, very little of the original ironworks remained; to the east of the road were a series of sheds and two kilns, covering almost all of the flat land available, and across the road on a platform created out of the old cinder heap were another group of sheds and four kilns. The buildings seem to have grown in a haphazard fashion, with kilns scattered about the site. The position was an awkward one, close to the clay and coal which the Maws had already been using at Worcester, but it was cramped and there was a steep drop down to the riverside, or by 1862 the railway line.

Despite the difficulties of the site, the time spent at Benthall was a period of innovation and success. After many experiments, commercial production of plain tiles, with many new colours, began in about 1856.[60] Enamel and majolica tiles were produced from 1862, the old plastic process was replaced by the dry, dust-pressed technique in 1863 and steam pressing was introduced into the existing layout in 1873.[61] But already the company was aware of the problems of the site, and from 1862 had been buying land for a new works on the river bank at Jackfield.[62]

Encaustic tiles were also being made in the 1860s at the old, run-down Jackfield pottery, where John Hawes had made earthenware.[63] Joined by Denny in 1867, they were replaced successively by Hargreaves and then Craven, and in 1872 the partnership of Craven Dunnill was formed, expressly to expand the manufacture of tiles, and to rebuild the old pottery as a new tileworks.[64]

Meanwhile, the first purpose-built encaustic tile manufactory had been constructed at Stoke in 1869.[65] The need for a new building was precipitated by a legal case between the china- and tile-making sides of the Minton Hollins business, and funded by the resulting compensation. Michael Daintry Hollins chose local architect Charles Lynam, and together they created a showpiece factory which combined the ornate façade of a pot bank with the principle of linear production, so that clay and products could be moved around the works in an orderly fashion.

For their new factory at Jackfield, completed only five years later in 1874, the Craven Dunnill partners chose the same architect. Four acres of ground were leased next to the pottery alongside the newly opened Severn Valley Railway and work commenced on the new buildings. The old works were later demolished,[66] and warehouses, showroom office and entrance lodge erected on the site. Lynam used some of the elements of his public buildings in Stoke such as the pointed arched windows and intricate ceramic detailing of the North Staffordshire Infirmary (1863)[67] in the façade of Craven Dunnill, and added an extraordinary spire (fig. 5.7). As at Stoke, clay was brought into the rear of the works, into a blunging house, and from there to slip kilns and mill room, presumably all in the eastern end of the site. It was then conveyed to purpose-built clay arks, where it was stored. The press shops, encaustic rooms, drying stove, saggar house, firing and glaze kilns seem to have been concentrated initially in the northern part of the building, culminating in the packing room, from which they were dispatched onto the railway. The tiles thus undertook a complete circuit of the works. There was a steam engine attached to the blunger, and a gas works at the rear of the site.

In 1879, construction was well under way at the new Maws factory for the manufacture of 'classical adjuncts of architectural embellishments, as well as of their

5.7 The Craven Dunnill Encaustic Tileworks at Jackfield, *c.*1878 (IGMT1983.1063)

porcelainic and fayence adaptations to articles of domestic utility' (fig. 5.8).[68] The new works finally opened in 1883. It was much larger than the Benthall site, covering $5\frac{1}{2}$ acres, and once again was designed to facilitate the movement of clay and tiles around the different processes. The most modern equipment, most of it designed by the company, was incorporated in special buildings, and provision made for extending the various departments without losing the relationship between premises. Unlike the other works, Maws included a fully mechanised firebrick works.

The three works illustrate changing ideas about tile production between 1869 and 1883. While the buildings at Stoke are very scattered, with wide carriage drives between each, both Maws and Craven Dunnill were on linear, cramped sites, and there is little space between buildings. At all three sites, the initial clay-preparation block remained separate, but while the workshops at Stoke are an isolated block, elsewhere they are much closer to the kilns, presumably so that ware can travel backwards and forwards as necessary. Stoke combined plastic and dust-pressed manufacture in different parts of the workshop block, whilst Maws was largely given over to dust-pressed techniques.

Power at Stoke[69] and at Craven Dunnill was initially limited to the separate mill block, whilst at Maws steam pressing was incorporated into the design of the workshops, as well as in the clay-preparation area, and the waste steam used to heat drying sheds. Craven Dunnill had its own gas-works in 1874.

The workshops at Stoke were heated by steam pipes and their own stoves; the 'green' tiles were then transferred to a hothouse or drying shed parallel to the kilns.

5.8 Maws tileworks at Jackfield *c.*1915 (IGMT1982.1740)

The sorting house and painters' shops were in a separate block with the finishing kilns, whereas at Maws the kilns were in a continuous line.

The three works also had very different transport arrangements; at Stoke there was no direct link either to the canal or to the railway; Craven Dunnill was built beside the railway, and Maws also had a separate railway siding to which goods were delivered directly. Maws had an additional tramway link to the coal and clay pits from which coal could be discharged into a subterranean area near to the kilns.

As with any factory site, the efficiency of the initial layout was short-lived. The most frequently altered and rebuilt element at each site was the kiln bank – at Stoke there were about 12 kilns, but by 1877 there were 20. At Craven Dunnill the up-draught kilns were converted to down-draught ones,[70] and a continuous gas-fired Dressler tunnel kiln was added in the middle of the site. Only five years after it opened, Maws were considering adding new kilns and equipment to their site. New gas plant was installed by the Horsehay Company, and the first firing of a new gas kiln using the plant was in 1905. A Dressler kiln was installed in 1919 outside the works to the north, which must have broken up any existing pattern of circulation.[71] Only 30 years after the new factory was completed, Craven Dunnill said that if the business was to be continued it would necessitate a general rearrangement of the works and a heavy outlay.[72] Maws continued operation, although in a much reduced form, until 1969, whereas at Craven Dunnill production wound down much earlier, and by 1952 a firm of precision engineers had taken over the site.

Tile-making is an example of a classic cycle of industrialisation which can be traced through the industrial works: a phase of experimentation as part of another business; the first larger operations on old or converted sites; and finally the confidence to invest in new efficient premises whose image, in turn, becomes part of marketing the product. Once built, these premises soon become cramped, circulation is interrupted, products change and production becomes inefficient, and the whole cycle may begin again elsewhere. What differentiated tileworks from brickworks or potteries was the sheer complexity of the manufacturing processes, many of which required specialist buildings. But such premises were expensive – although the first Stoke factory was built on the basis of a compensation claim, Craven Dunnill and Maws were built for a more buoyant market, by well-capitalised partnerships. It is difficult to see how the expenditure could be justified in an out-of-the-way area, without the confidence afforded by a national, still less an international, market.

Adaptive re-use: industrial buildings as an economic resource

Cheap and flexible premises remaining after one industry after another had closed in the early twentieth century actually acted as an incentive to new industries to move into the area. Instead of the area stagnating, there were probably more new small industries at this time than at any other period, and thus the industrial cycle was re-established.

In fact the adaptive reuse of industrial buildings was an old-established practice.[73] If they hadn't been shaken to pieces, ironworking sheds were particularly adaptable – two of the sheds at Newdale became a barn and a meeting house in the 1760s, and the Upper Forge became a nineteenth-century mill. Even specific structures such as iron furnaces were reused – Abraham Darby adapted an old charcoal furnace, and at Bedlam an iron furnace became a brick kiln. There was by no means a firm separation between industrial and residential buildings – the conversion of redundant ironworking buildings into tenements, often while the rest of the site was still in operation, occurs frequently in Coalbrookdale in the early nineteenth century[74] and can be used as a measure of declining industrial activity. Mining buildings survive less often, although at least one mining engine house at Jackfield is now a house.

Old buildings fulfilled a much more important function – that of attracting new industries, particularly to settled areas, where there was a demand for goods and services by the local working population. The two building types commonly found in urban areas, which proved most useful, were malthouses and warehouses.

Malthouses were one of the most common urban building types in industrial use. In the late eighteenth century, most pubs brewed their own beer, but with the centralisation of brewing in the hands of larger firms local brewing was abandoned. Attached to most local public houses of this date is a malthouse, usually a two-storeyed gabled building with a large open space inside, tiny windows to restrict light, and a kiln at one end, providing a useful space that could be readily adapted to industrial needs. Local malthouses were converted to all sorts of industrial purposes – a bakery on the Wharfage, a cinema at Dale End, an aluminium smelter at Madeley. Because the structure of a malthouse was so close to that of an industrial shed, malthouses themselves were often adapted from industrial buildings – Brooke's steelhouse was a malthouse by 1734, and the Benthall lead smelter had been converted by 1765.

On the Ironbridge Wharfage during the nineteenth century were sieve- and riddle-makers, wiredrawers, blacksmiths, tinsmiths, ironmongers, and the makers of iron patten rings for workmen's clogs.[75] Many of these were located either in malthouses or, more often, in the old two-storey warehouses built to serve the river trade, which were no longer needed after the construction of the railways. Such industries served the needs of a working population, and in a sense were the logical precursors of the industries which today most commonly occupy old industrial buildings – garages.

As significant users of old buildings, garages have been among the most important influences in the conservation of historic industrial buildings. The earliest garages appear in Ironbridge in about 1922, and by 1941 there were four motor engineers in Coalbrookdale, and several more on the Wharfage and at Dale End, almost certainly in existing buildings. At Coalbrookdale, Lower Forge is still occupied as a garage, and elsewhere the Benthall pipeworks and the Coalford brick-drying shed are used for truck repairs. Similar businesses occupied the Coalbrookdale Upper Forge (plant hire firm) and the Blists Hill brickworks (haulage contractor) in the 1950s and 1960s.

The King Street Tobacco Pipeworks in Broseley provides an example of the way in which a larger business – in this case a small factory – was able to move into and adapt an old warehouse. On a patch of common land in Broseley Wood, a long timber-framed building was constructed, with a small vented brick stable added at the north end. The long building was demolished to its base wall level, and reconstructed as a long brick building, with an open hoistway facing the street, and six enormous windows in the rear wall. By 1838, the big windows had been blocked for a seed warehouse with two tenements on the ground floor, and the vented stable block was now offices. Later the warehouse was converted to two more tenements and a pair of cottages was built at the back of the building.

The site only became a pipeworks relatively late in its history, when it was bought in 1883 by Rowland Smitheman, a local builder, who converted it into a factory. Pipemaking, like china manufacture, required workshops rather than large open sheds, and it was a simple matter to convert warehouses and cottages into the necessary accommodation. A kiln was erected in the courtyard, the cottages knocked through into one building, and the upper storeys of the warehouse opened up again. The old office became a chaff store and packing area, and a new office was set up in one of the tenements on the ground floor. As Broseley Wood became built up and access more difficult, the old entrance to the north was blocked, and big double doors inserted. Smitheman sold the site to the Southorn family who already made clay pipes elsewhere in Broseley, and who kept it in operation until the late 1950s, when it was abandoned with the equipment still intact.

Old buildings ideally served the needs of such small workshop enterprises, typical of most built-up areas. On the whole they were small craft-based industries directly servicing local needs with low capital investment and a high element of skill. They were individual operations, and did not rely on any organised putting-out system.

Nevertheless, there was also a tradition of small, purpose-built industrial workshops, often located in urban areas amongst housing. These were usually small sheds, on a domestic scale, but they often displayed features characteristic of a particular industry such as hearths or ventilation systems. They are common in the West Midlands, where smithing, chainmaking, weaving, leatherworking and nailing each produced a characteristic building type.[76] One of the main problems with such buildings is survival – the rate of destruction of outbuildings tends to be much higher than street-front buildings as rear yards are converted into gardens or outbuildings into unrecognisable garages. A good group of such buildings remains in Much Wenlock,[77] but relatively few have survived in the Gorge. This in itself makes the Old Armoury in Ironbridge – one of the few purpose-built workshops to survive – particularly unusual. High up on the hillside, the building originally had two storeys at the back and one facing the street. At the lower level, heavy brick and iron arches support a first-floor working area of seven hearths set around the walls, lit by windows, which were, however, truncated by the addition of the third storey. The windows provided direct light to each hearth, implying a level of fine and detailed work. Used as assembly rooms in the late nineteenth century, the stone chimneys of the hearths can still be traced as scars in the wall. The building was used by a patten-ring maker (manufacturing the iron rings on the bases of workers' clogs) in 1847, but may originally have been built for a forging business such as Walton's, which made chains in Ironbridge.

One small industrial building also stands in Church Street. It is a simple two-storey structure, entered from the upper floor, with evidence for a hearth in the basement. Dating to the early nineteenth century, it remains in use for light industry and metalworking today.

Pioneer industries

The closure of the large traditional industries left a supply of abandoned industrial complexes, as well as an unemployed skilled labour force. These conditions, in the context of the government planning policies of the two world wars, attracted a series of new industries to the area. Such enterprises had no links with the mining, ironworking or ceramics industries which had so long dominated the local economy; they were small-scale, flexible enterprises, responding to the changing demands of consumers and often pioneering new products. The Gorge had, in effect, become a nursery for new industry.

The old Coalport factory was a particularly useful site. The Nuway Manufacturing Company had moved in there in 1927, and in 1941, after heavy bombing, the London firm of Chillcots moved their sheet-metal Gasket Division into part of the old works, where they made gaskets, washers and exhaust systems which were stacked in the kilns. In another part of the works Frank Hawker Ltd made fancy metalware powder compacts, lipsticks, mirrors and trays. The company had moved in in the 1940s, at the request of the Ministry of Supply, and operated until the mid-1960s.[78]

Other firms moved into the area to start new enterprises. In the Coalbrookdale Company's Severn Foundry, an employee of a Wellington toy firm set up the Merrythought Company to manufacture teddy-bears and soft toys. The building was light but a little too large; it was easily adapted, however, by being subdivided into sewing and cutting areas. Teddy-bears are still made in Ironbridge today, but many more of the industries which began life in old buildings in the Gorge have since moved away into larger premises. In the old Drill Hall, Clifford Williams began making pyjamas and khaki shorts in 1951 and when the space proved too small the firm acquired the Old Armoury for storage. It was so successful that in 1962 a new purpose-built works was built in Queen Street, Madeley, and since then plants have been built in Telford, Leominster and Oswestry. A similar process of expansion from a small pioneering industry in old buildings occurred at the old Calcutts Ironworks, where W.H. Smith began making machinery for the local clay industry in the 1870s, as well as general smithing, iron and brass founding. The firm made aeroplane parts during World War II and in 1941 were taken over by Messrs Marshall Osborne & Co. Ten years later they took over the old Craven Dunnill tileworks, and since then have moved into new premises in Telford.

Industry today still follows the same pattern of small firms in old industrial premises, especially on the south bank of the river, outside the area of Telford. Here copper reclamation takes place at the old Rock Tileworks, gates are manufactured on the old Milburgh Tiles site, and garden sheds assembled in one of the old Doughty Tileworks buildings. The old Maws works now houses craft industries. On the north side of the river, furnace heating elements are made in the Old Armoury in Ironbridge, rubber products are made at part of the Coalport works and teddy-bears at the old Coalbrookdale foundry.

Thus the supply of old industrial buildings played a central role in continuing the industrial cycle by attracting new industries, which, once established and successful, would move out into larger premises. In fact, it might be suggested that the opportunity to establish a brand-new, purpose-built works is more the exception than the rule in the general course of industrialisation. This pattern is well established for other parts of the country too.[79]

Archaeology, architecture and functionalism in industrial buildings

This study has taken an explicitly archaeological view of industrial buildings, concentrating upon the physical evidence for the buildings themselves, their setting and, where possible, their development through time. Inevitably, it is a study which lacks personalities or finances and may not satisfy economic historians but it nevertheless provides an opportunity to explore the nature of the physical evidence – and, by implication, the importance of buildings themselves, however humble, as a resource.

The history of industrial buildings has most often been expressed in terms of machines and power, particularly in the application of centralised power to the textile mill. But here power is far more important in the landscape than in building development. As often as not, power was bodged into buildings, well after their construction – a hammer engine added to the Upper Forge, or steam added to an existing brickworks. A multistorey porcelain workshop may superficially resemble a textile mill, but the power was applied elsewhere, and even away from the site. The concept of centralised power is therefore rarely helpful in understanding the original plan or structure of local buildings, although it is often a major contributing factor to their evolution.

If power is only one of many factors in a building's evolution, how relevant is the concept of the factory in the Gorge? The legislation relating to factories referred almost exclusively to textile factories until well into the nineteenth century, and contemporary writers excluded organisations in which 'mechanisms do not form a connected series, nor are dependent on one prime mover';[80] more recently the concept has been widened to encompass anywhere where the means of production are concentrated on one site.[81] But by becoming more general the concept has become less useful. It seems more appropriate to refer to 'industrial complexes', a term which better encompasses the variety of building types which might be found on any one site, and to look for other models in order to understand individual buildings.

Equally, the term 'factory' does not easily fit one of the industrial building types – the ironworking shed where the aim is not so much to accommodate people and power as to protect, support and ventilate large, space-hungry processes. With a few exceptions (Soho), early ironworking buildings have often been derided as grim and austere,[82] their form subservient to any process which may go on inside. Yet it is possible to see a generalised single-storey shed, developing as a building type in its own right, irrespective of the processes it contains. In the Gorge the use of bricks and of gable parapets places its origins firmly in the local tradition of barns and of small industrial housing. But the demands of strength and light create new features – timber trusses with iron ties, the use of iron to support buildings and for light, the increasingly fine, multipaned cast-iron window. Ventilation comes through roof vents, or the typical circular gable opening and sweeping arches provide ingress for bulky objects.

Is it more useful, then, to look at these building types in terms of the debate between functionalism and architecture? As a style, the former is dominated by functional considerations, and acquires aesthetic value almost coincidentally. But the important factors in the industrial shed are space, light and ventilation, and these could be achieved as easily by a roof propped up by pillars. In each of the ironworking sheds there is a clear attempt at an aesthetic – arched multipaned windows where shutters might have done, decorative brickwork, gable details, etc. Yet this is not an aesthetic which relates to major contemporary architectural trends. Its roots are instead to be found in local vernacular practice, and it continues to develop in the nineteenth century, not with reference to domestic structures, but by now established industrial motifs. Multipaned windows, brickwork, roofs continue to evolve along their own trajectory as an industrial mode of practice, but until more early ironworking buildings are identified, the buildings of the Gorge must stand as a model for the origins of a coherent industrial building type.

As well as setting at least one industrial building type firmly within local practice and pragmatism, it has been possible to look at complexes or groups of buildings as working units. This is best done through ceramics, where a startling array of buildings and forms is geared towards one basic process – that of preparing and burning clay. The degree of complexity is not simply governed by technology (it was over 50 years before innovations in pottery-making were applied to brick), but by external, and to archaeologists invisible, factors, such as markets and capital. Within the same industry, the buildings document very different strategies which at their most basic must relate to profit – if sufficient value could be added to the clay, the outlay on works was justified. Once again, it is not possible to understand buildings purely in terms of the application of the most up-to-date and efficient technology: they have to be seen much more as products of a local – and economic – context.

But a longer archaeological view also makes it clear that any advantage to be gained from a purpose-built works, or scheme of efficiency, was short-lived. And this underlines one of the eternal problems in establishing the pattern of flow in a ceramics operation. Kilns in particular were frequently rebuilt on a 20- or 30-year cycle, and equipment was changed to meet market demands. Most processes left few marks on the surviving buildings, and without documentation the task is a daunting one. However a works was initially conceived, it was unlikely to stay that way for long.

Not only did the function of buildings change within industries, it also changed from industry to industry. One of the main fallacies of an architectural or a functional approach to industrial buildings is to treat them as static entities with a single purpose. The changing pattern of building use is as much an industrial trend here as the expansion of the iron industry or the growth of brickmaking. The small and mobile industries were actually encouraged by the availability of buildings, in the same way as raw materials attracted the larger industries. Buildings in this case become more than a product of industrialisation: they become a factor in their own right.

Finally, all of these buildings have to be seen within the peculiar landscape of the Gorge, against a recurring pattern whereby the initial attraction of raw materials inevitably had to be set against the high cost of transport, and the eventual exhaustion of minerals. However elaborate the works, the realities of day-to-day operation in a difficult valley with no large centre of population nearby had to be overcome. Buildings are not isolated entities, but functioning parts of the landscape which cannot be considered separately.

As in the industrial landscape, the two fundamental elements of an archaeological approach – context and change – come together to provide an alternative perspective on industrial buildings. The advantage of a small area, containing a very diverse sample of buildings, also acts as a discipline, making it less easy to select the biggest, the best or the most typical. There is much more to be learned from the buildings of the Gorge, especially by the detailed survey of the less obvious structures, but only if they survive the pressures of modern development, with their information reasonably intact.

6

LANDSCAPE OF HOUSING: THE GEOGRAPHY OF SETTLEMENT

'The village of Madeley, comprehending hundreds of detached houses, . . . reminded me of the drawings of a Chinese town, as the same indiscriminate jumble of beauties.'[1] This description, inspired by a popular view of Chinese art, was written shortly after the Iron Bridge had been built. The same idea of a Chinese landscape was used in a description of Cromford, and conveys striking similarities with other rural industrial landscapes, as for example the Pennine landscape of textile outworking described by Daniel Defoe, where there were 'Houses scattered and spread upon the sides of the hills even from the bottom to the top'.[2] In this apparently random scatter of settlement the landscape of the Gorge is connected with other areas where industry developed outside the structure of urban or planned development.

To early visitors, the landscape of the Gorge was also an aesthetic landscape, perceived as an embodiment of sublime disorder. In this 'jumble of beauties', however, a clear pattern of land use is discernible, and the development of settlement can be closely connected to the changing economic organisation of land and resources in the three Gorge-side parishes of Madeley, Benthall and Broseley. This chapter studies the changing relationships both among settlements and between industry and settlement, using these themes to trace the development of the Gorge as an industrial area. It charts the gradual transformation of a 'rural conurbation' of separate hamlets with distinct economies to an integrated urban area centred on the small town of Ironbridge.

Origins of industrial settlement: before 1700

Before the dissolution of the monasteries, settlement on the slopes of the Gorge was confined to a number of farms in woodland clearances, an occasional mill, perhaps a small community in Coalbrookdale. There are isolated references to the exploitation of minerals in the Gorge since at least the fourteenth century, and there are references to mining and iron manufacture that date back even earlier than this (see chapter 2), but nothing to suggest that this was on any scale, or that it generated any permanent new settlement. The principal focus of settlement remained the nuclei of Benthall, Broseley and Madeley. Settlement in the Gorge itself was minimal, and has left little trace.[3]

But from the end of the sixteenth century the slopes of the Gorge became the focus for the growth of settlement on a new scale, at a time when the exploitation of its mineral resources was given serious attention by the landholders who had purchased the old monastic estate. There are references in manorial documents to new settlement in the parishes of Broseley and Benthall outside the old nuclei,[4] and after *c.*1620 there are a number of maps which trace the development of settlement in the area of Broseley Wood. [5] An observer in 1690 noted that 'the ancient waste of commons of Broseley is now in greatest measure built up and enclosed by poor people, and has become as a county town'.[6] By the end of the seventeenth century, references to unlicensed cottage-building in Madeley Wood make it clear that this area also was in the process of settlement, and the Easter Book for 1711 lists 148 names in Madeley Wood alone.[7] A series of probate inventories refers to settlement in the Gorge area after *c.*1675, giving some idea of the nature of its communities and their resources.[8]

In spite of the documentary evidence for settlement during the late sixteenth and early seventeenth centuries, the only earlier buildings to have survived are the small number of large houses built in the early seventeenth century by mine- and barge-owners. It is only from the later seventeenth century that there are significant numbers of small dwellings and cottages either still surviving or for which there is pictorial or photographic evidence. Perhaps it was only in this period that buildings were constructed in such a way that they were valued enough to survive. The houses of the late seventeenth century were built by professional building craftsmen – their predecessors may have been more primitive structures (see chapter 8). The emergence of what were to become permanent buildings marks a key stage in the development of the Gorge, and is an indicator of the resources generated in new communities dependent on mining and manufacturing. It is from this period that the development of settlement can be charted in some detail.[9]

The location of late-seventeenth-century settlement can be precisely mapped across the Gorge (fig. 6.1). The pattern of development suggests dispersed settlement ordered as a number of loose nuclei or small hamlets. On the north bank of the river, although some houses were built in relative isolation, most formed loose clusters of building, perhaps the largest being at Madeley Wood Green, where Bedlam Hall was built in the early seventeenth century, and where there were several cottages.[10] Some of these small settled areas can be related to specific patterns of economic activity – clusters of early houses near what was to become the site of the Iron Bridge and at the Lloyds mark wharfage areas and ferry crossings at pools on the river, and on the slopes of the Gorge early settlement closely followed the outcropping coal seams (fig. 6.2).

These hamlets had commonly known identities and boundaries, and their differentiation remained important into the twentieth century. The nineteenth-century census enumerators, for example, were confident of the difference between Foxholes, Brockholes, and Lincoln Hill in Ironbridge, or between Coalford, Lloyds Head, Calcutts, Jackfield and Salthouses on the south bank. These different local names recalled the several separate communities formed in the early period of settlement in the Gorge.

South of the river, settlement followed a similar pattern. In Benthall, the medieval

6.1 The beginnings of settlement: Coalbrookdale and Madeley Wood before 1700. Map based on surviving and documented early buildings (N. Smith)

village had been deserted by the mid-seventeenth century, and encouragement was given to settlement for mining as the outcropping coal measures in the east of the parish were exploited.[11] The earliest surviving buildings are alongside areas where the remains of bell pits provide evidence of early mining activity. Similarly, in Broseley, new settlements were established at this time on the former commons of the parish, distinct from the nucleated main settlement, and again connected with the development of mining – the earliest recorded new settlement was at Coalpit Heath, where a coalpit and associated houses were recorded as early as 1550.[12]

A series of separate settlements along the river were also linked with mining, the domestic pottery industry and with river traffic: Coalford, Salthouses and Lloyds Head all had substantial numbers of seventeenth-century buildings, and one late-seventeenth-century house still survives at Calcutts – a community at Ladywood, and another at Jackfield are shown on the map of 1620. Calcutts and Ladywood were both loading-points for the transport of coal from nearby mines.[13]

These new settlements had a distinctive form. The individual houses were for the most part detached in small plots of land, but were concentrated in particular areas.[14] A number of large and ornately constructed buildings were the focus of several of these hamlets: two such houses were built in the early seventeenth century at Coalford,

6.2(a) The Lloyds, Madeley Wood, *c.*1880–1900. A small settlement established at a river crossing by the early seventeenth century served as a focus for successive generations of development, the tight cluster of cottages around Lloyds Hall contrasting with the scatter of houses beyond (IGMT 1988.284)

(b) Severnside, Ironbridge, 1895–1905. The tight packing of houses below the Iron Bridge suggest the importance of this site, which had been a river crossing long before the building of the bridge (IGMT 1982.1863)

and they faced another across the river (Lloyds Hall). These houses were all timber-framed with a two-storeyed hall and cross-wing plan form, the façades embellished by a series of gables marking the three principal bays. In 1672, the owner of one of these houses at Coalford was Adam Crompton, a master collier. Another substantial house, Bedlam Hall, was described as a Jacobean mansion before it was demolished in the nineteenth century.

Other houses which survived from this period were considerably smaller – most were cottages with a two-unit plan, though Parsons' map of Broseley in *c.*1620 suggests that there were also a number of single-celled dwellings. The profile of housing which can be derived from Hearth Tax returns and probate inventories, while weighted towards the better-off members of the community, confirms this pattern of building distribution: large numbers of small cottages, and a small number of more substantial dwellings.[15] This housing profile corresponds closely with what is known about the organisation of the early industrial society – the division between master colliers and labourers is mirrored in the structure of housing stock: it was the master colliers who were responsible for building the larger houses; the rest were the cottages of labourers.[16]

There was a coherence in the structure of these early settlements that suggests that they were neither marginal nor poor. Studies using probate inventories and other documentary sources have demonstrated the presence of general traders in these areas by the late seventeenth century: the Gorge was not an entrepot for a region – these traders served the local population. Their presence, and the growing importance of consumption within the home, give an index of prosperity and suggest the significance of a cash income for the population of the Gorge.[17]

Documentary and map evidence suggests that these seventeenth-century settlements were primary, the first major development of this land for building. Although mining and manufacturing are recorded for a much earlier period in both Benthall and Madeley Wood, there is nothing to suggest that these activities had generated any permanent settlement. The absence of a settlement record (and of evidence of coherent and sustained industrial activity) may suggest that this early industry was small in scale, probably seasonal, perhaps temporary. This accords with the early development of industry in other areas. In South Wales, for example, the early charcoal iron furnaces relied on seasonal work – housing was provided only for key workers, and labourers may have squatted or lodged in farmhouses.[18]

Neither is there any evidence that these were marginal agricultural settlements turning to industry as a means of eking out a poor livelihood from the land. Documentary records show that the first settlers in Benthall and the outlying parts of Broseley were identified as miners; they were also described as incomers to the area.[19] While probate inventories show that it was not uncommon for miners to keep some stock, and the distribution of houses in plots of land enabled small-scale cultivation and animal husbandry, it is clear that such activities were secondary in the economy of the area, and the plots surrounding the houses were too small to sustain anything other than the simplest husbandry.[20]

But the new settlements were not established on unused or waste land. The northern slopes of the Gorge comprised part of the woodland of the parish of Madeley and were an integral part of its pre-industrial economic structure, used for coppice timber and for pannage. But a process of enclosure associated with the growing importance of pastoral farming had already begun to reduce the importance of wood and arable in the agrarian economy of the parish well before the seventeenth century, and by this time mining was already under way on the slopes of the Gorge.[21] Unsuited for cultivation, the wooded slopes of the Gorge were the principal locus of early mining, and also provided fuel for the charcoal iron furnaces.

A similar pattern of land use can be identified in Benthall, where there was a more radical shift in its organisation as the medieval agricultural village was literally replaced by a new industrial settlement on the margins of the parish. The estate was simultaneously exploited for agriculture, and the woods of Benthall Edge were an integral part of its economy.[22]

In Coalbrookdale the land formed part of a number of farms already independent of the manor of Madeley by the seventeenth century. These were progressively sold in small lots in the early years of the eighteenth century. Some of the land in the Dale remained in agricultural use until the mid-nineteenth century, and housing crowded the steeper slopes which could not be productively used for agriculture.[23]

In Broseley, the greater part of the industrial settlement (the area known as Broseley Wood) had previously formed part of the common lands of the parish, and the early settlement of this area sparked off riots within the parish, the villagers jealous of their lost common rights.[24] But this was not a straightforward conflict between agricultural and industrial interests. Large blocks of common and open fields were steadily eroded by enclosure and the formation of consolidated holdings during the seventeenth century, a process illustrated in the series of estate maps showing the formation of individual tenements taken from the former common.[25] Landowners may even have encouraged building on the commons as a means of consolidating their rights over the mineral deposits. Settlement associated with mining therefore appears to have been part of a wider reorganisation of land use within the parish.

To the north a number of large farms (probably originating as woodland clearances) were already separate from the principal part of the parish by the early seventeenth century, and it was on part of these estates that most riverside development took place.[26] This development formed a series of small separate settlements which closely mirror the pattern of landholding on these farms.

Changes in the organisation of the economy are underscored in the roles taken by landlords in the exploitation of mineral resources. In all three parishes, the stimulus to seventeenth-century exploitation of minerals came from the major landholders, who made a substantial investment in mining and transport networks.[27]

Industrial settlement can therefore be placed in the context of more general shifts in the economic organisation of the parishes of the Gorge. These shifts encouraged the more rapid development of mining and manufacturing from the mid-seventeenth century as wealth was invested in the exploitation of resources on the new estates created by the dissolution of Wenlock Priory. At the same time, the pattern of land use was modified by the creation of consolidated holdings. Enclosure is sometimes said to have entailed impoverishment, the creation of surplus labour and under-employment;[28] close links between the enclosed landscapes of pastoral farming and rural industry have been demonstrated elsewhere. While there is no evidence for an impoverished rural population in the Gorge turning to industry to eke out meagre agricultural incomes, the loss of common rights certainly brought about a change in the basis of rural labour and may have been an important precondition for the creation of an industrial labour force. In addition, the formation of small consolidated holdings perhaps encouraged the development of alternative sources of income from under-used or unprofitable land, either by industry or housing development.

Industrialisation: the eighteenth century

The early eighteenth century witnessed a major phase of building and expansion of settlement throughout the Gorge, alongside the continued exploitation of its resources for mining and manufacturing. Many buildings associated with this phase of growth survive and their record is augmented by a series of descriptions and illustrations of the Gorge as it became a notable place to visit later in the century. In addition, a documentary record provides substantial evidence for the process of settlement in title deeds and probate inventories.[29]

Eighteenth-century settlement consolidated development in those areas where building had begun in the previous century. The limits of these old settled areas were extended, and the density of their development increased. Distinctive clusters of building can be identified as small terraces and groups of building were created through a process of extension, subdivision and new building (fig. 6.3) (see chapter 7).[30]

Some of these clusters formed substantial nuclei that were already crowded with building by the middle of the century – one such nucleus in Madeley Wood was centred on the Golden Ball Inn which was first licensed in 1728. Another nucleus was at Loadcroft Wharf, where by mid-century there was a long terrace of 10 dwellings later known as Nailers Row, at least nine other houses in a tightly built group, two houses on the site of the Swan Inn, and two further terraces of around six and seven dwellings built up along the line of Lincoln Hill, which was already an established route in and out of the Gorge.[31]

In other areas, development was more dispersed – in the Brockholes and Foxholes, for example, cottages were strung out along the slopes of the Gorge, following the outcropping coal seams. But in all areas it was now common to find individual plots increasingly crowded, as new development concentrated in particular areas, rather than a uniform spread of building across the landscape.

Many of the features of the eighteenth-century landscape of settlement can be illustrated by the development of Coalbrookdale:

> In the year one thousand seven hundred, the whole village consisted of only one furnace, five dwelling houses and a forge or two . . . [now] its trade and buildings are so far increased that it contains at least 450 inhabitants and finds employment for 800 people.

6.3 The impact of the industrialisation: Coalbrookdale and Madeley Wood, *c.*1750. Map based on surviving and documented buildings (N. Smith)

This was how George Perry described the development of Coalbrookdale in 1753.[32] This pattern of growth is illustrated in the engravings to which this text was an accompaniment: they show two distinct settlement types in upper Coalbrookdale – on the one hand, a pattern of single cottages dispersed in small plots; on the other, the earliest terraced housing.[33] Surviving buildings show that it was the terraced housing which became characteristic of Coalbrookdale's development – the single cottages were gradually replaced by other buildings. The engravings record some ten houses in addition to the six dwellings of Tea Kettle Row (fig. 6.4). Also at the upper end of the Dale by this time there were at least ten other houses in three short rows – other terraces were built close to the Upper and Lower Forge sites.[34]

The development of Coalbrookdale was also marked by the new building of substantial houses for the ironmasters and managers of the ironworks. The earliest of these were close to the principal site of the ironworks, and their development included the creation of a formal landscaped garden: Dale House was built in 1715, Rosehill in 1720 for Richard Ford, manager of the ironworks. By mid-century, the Darby family had built another larger house some distance away (Sunnyside, built in 1750).[35]

6.4 'The Upper Works at Coalbrookdale', 1758. A scatter of cottages loosely grouped around the works, the larger houses of the ironmasters and their managers above the furnace pool, with Tea Kettle Row, the earliest recorded terrace in the Gorge, behind (F. Vivares, Elton Collection: IGMT)

There was also a small number of houses of an intermediate size in the upper dale which tradition links with the management of the ironworks.[36]

A similar pattern emerged elsewhere in the Gorge. By the late 1750s, there were at least six substantial houses to the north of the river: Belmont House was built on Hodgebower in 1753, in an area being developed with small rows and groups of cottages. Two substantial houses by the river may be of a similar date, and Severn House was built *c.*1757 by George Goodwin, a master collier and partner in the Madeley Wood Company.[37] On the south bank, Calcutts House was built in 1755 alongside cottage rows.[38] None of these houses was a farm, and none had the basis for substantial landed estates: Dale House in Coalbrookdale was built on a plot of land that was only 60 yards by 60 yards in 1715 – the Darby family consolidated landholding in this area only slowly over the next century and a half.[39]

The clear hierarchy of housing types shows both the social structure and the organisation of the industrial settlements. Industrial work was not a marginal occupation associated with poverty, but had generated its own organisational structure, concomitant with its expansion and increasing capitalisation. It was agriculture which continued as a marginal activity. During the eighteenth century, the small clusters of houses that had characterised the early industrial settlements gradually became more complex groupings with distinct identities. There was a shift from an 'industrialised countryside'[40] to a series of small industrial villages, a kind of rural conurbation, quite distinct from the principal townships of Madeley and Broseley, each with its own economic and social structure clearly reflected in the number and range of building types.

Broseley Wood represents an exception to this pattern. It was the largest area of development which was predominantly workers' housing in the Gorge. The settlement was said to have been built up by poor people, 'the dregs of many counties'.[41] During the eighteenth century, industrial wealth was invested in building in Broseley itself, which rose to pre-eminence and came to be important as a market-town. Many of its eighteenth-century town houses survive, and can be directly compared with the larger houses within the Gorge. Broseley Wood remained an integral part of Broseley itself and formed an early industrial suburb of labour within a stratified urban spatial and social structure.[42]

Settlement and the organisation of industry

The pattern of settlement that had emerged in the Gorge by the mid-eighteenth century was an integral part of a wider system of land use and ownership. Settlement patterns form an important part of the evidence in a reconstruction of the process of industrialisation, and the organisation and structure of industry.

One crucial determinant of settlement location was the pattern of landholding underpinning all activity on the land (see chapter 7).[43] But settlement location also bore a strong link with the availability of raw materials, especially in the early phases of industrial development. In Benthall the first industrial houses were in close proximity to the bell pits of early mining activity, and in Madeley Wood there is a clear connection between early settlement and the outcropping coal seams. In addition, the development of more specialised mining was linked with the involvement of landowners, providing the capital (in mining itself and also in transport) and making available the land (see chapter 2). The particular disposition of settlement reflected the existence of mining freeholds or leaseholds, and there is a close correlation between the pattern of landholding and the location of settlement which is particularly clear in the spread of communities along the south bank of the river.

The closeness of the links between the location of manufacture and its associated settlement is also borne out in Coalbrookdale, where the main areas of settlement were immediately adjacent to the furnaces and forges. Similarly, settlement clusters by the river map out wharfage areas and ferry sites: on the south bank, some of these were originally associated with mining, developing especially as loading points for the shipment of coal. These settlements also made use of other local resources in the establishment of a domestic pottery and clay-pipe industry. At Jackfield, small kilns were attached to cottage rows (fig. 6.5a–b).[44] The importance of the river traffic was signalled in the nature of housing: it also generated a domestic economy of inns, boarding-houses and brothels. Riverside inns were among the largest of the buildings alongside the river.

But the links between housing and industry go beyond the simple matter of

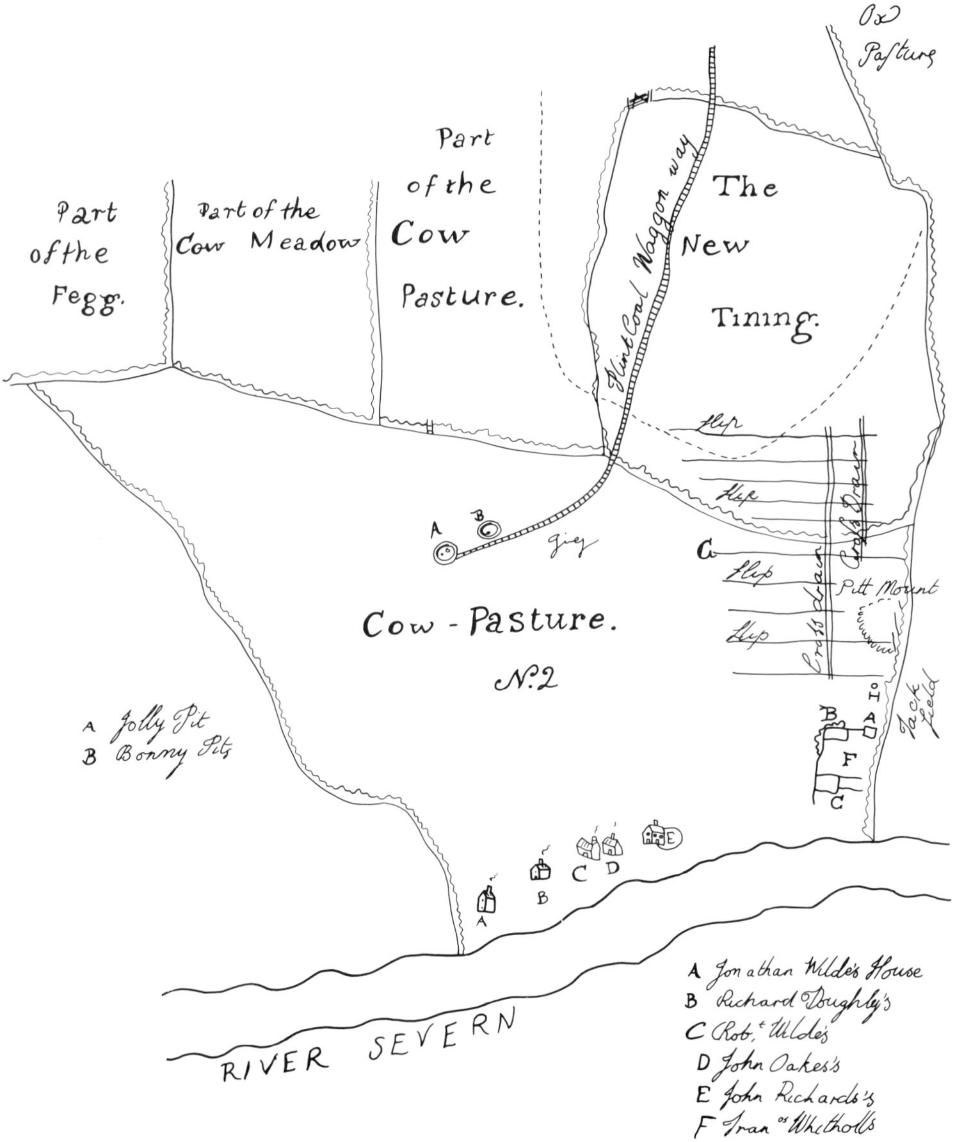

6.5(*a*) Jackfield and (*b*) Calcutts *c.*1720. The Broseley Hall Estate Book records several distinct settlements on the south bank of the river. The community at Jackfield had its *raison d'être* in pottery production.

location. The organisation and structure of industry is closely reflected in the organisation of housing – this can be demonstrated by investigation of the connections between changes in industrial organisation and changes in the form of settlement, in particular in the case of mining.

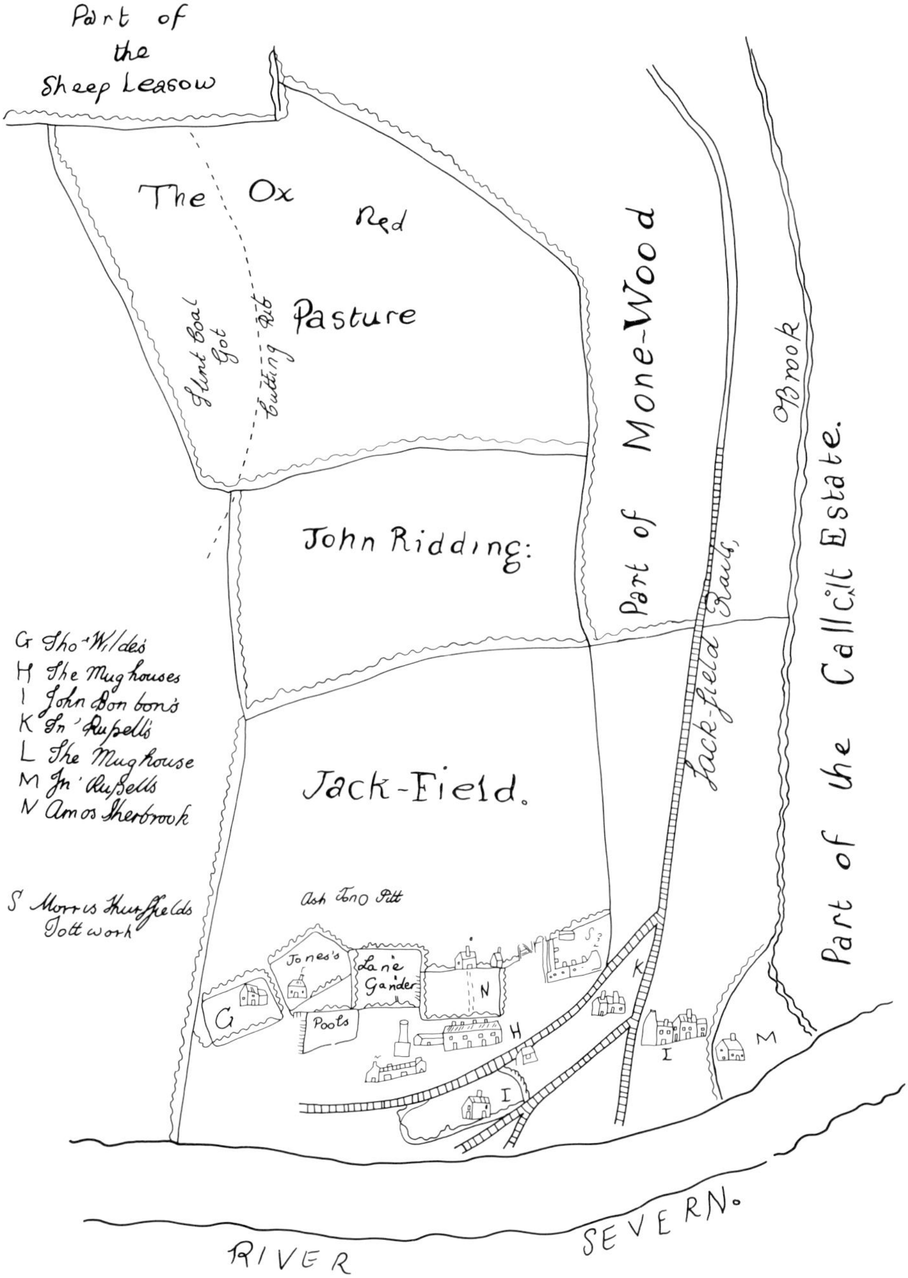
Part of the Sheep Leasow
The Ox
Red
Pasture
John Ridding:
Jack-Field.
Part of Mone-Wood
Brook
Part of the Callcott Estate.
Jack-field Rails
G Tho. Wildes
H The Mughouses
I John Bonbon's
K Jn. Russells
L The Mughouse
M Jn. Russells
N Amos Sherbrook
Jones's
Lane Gander
Pools
RIVER SEVERN.

(*b*)

The development of mining in the area can be traced from the relatively simple bell-pit mining through to long-wall adit mining and deep pit mining. The shift in mining technologies appears to have taken place at different times in the different areas of the Gorge – in Broseley, the long-wall method had already been adopted by the mid-seventeenth century and surface extraction given place to deeper pit or adit mining. In Benthall, mining was still by bell pit in the later seventeenth century. In both cases, mining was a primary occupation from the outset.[45]

The greater investment required by more complex mining systems was associated with the employment of a more specialised labour force. In the Black Country, the development of deep mining techniques was associated with the introduction of more specialised work, and with the decline of patterns of by-employment which had been linked with small-scale, bell-pit mining. New techniques demanded new levels of skill and an increasingly capitalised industry required a committed, full-time workforce.[46]

In the Gorge, settlement was associated with mining and manufacturing from the outset: the seventeenth-century settlers of Benthall and Broseley Wood were described as miners,[47] and although the small enclosures associated with the houses would have sustained small-scale cultivation and animal husbandry the plots were not big enough for subsistence agriculture. These plots were progressively reduced in size as more housing was built during the eighteenth century.[48]

Housing and the mines themselves were in close proximity. This close physical association may itself reflect the scale of industry and its organisation: in south-west Lancashire, for example, permanent housing was a specific product of an increasing scale of production.[49] Survival of any settlement at all gives an indication of the importance of industrial work.

The permanence of the houses themselves, especially from the late seventeenth century, also suggests the importance of money in the economy. Little survives from the earliest phases of documented industrial exploitation and associated settlement (late sixteenth century to early seventeenth century), suggesting that a first generation of housing was less substantial, to be replaced as soon as sufficient capital had been acquired from wage labour. Fragmentary remains of early building phases incorporated into eighteenth-century cottages give physical evidence for a process of rebuilding.[50] The largest houses of the Gorge that were built in the seventeenth century appear to represent the investment of wealth derived from mining. It is only from the late seventeenth century, and especially from the early eighteenth, that workers' cottages survive in any numbers.

The separation of industrial work from that of agriculture also influenced the nature of the houses built. In Benthall, late-seventeenth-century houses are generally larger than their successors: this gives an indication not merely of investment but also of economic organisation. The size of these earliest houses suggests an economy in which small-scale agriculture played a part: this would have necessitated additional domestic service space for the processing and storage of produce. An economy dominated by specialised and capital-intensive industrial work taking place largely outside the home and reliant on markets rather than subsistence agriculture permitted dwellings of a smaller size.[51] Cottages built in Benthall during the eighteenth century were often very small, and subdivisions of earlier larger houses over the same period provide further evidence for changes in the use of domestic space.[52]

Housing development in Madeley Wood followed a similar pattern. A number of detached houses built in the later seventeenth century were subdivided and added to in the early eighteenth century to create rows and clusters of very small dwellings. At the same time, several single-celled houses were built.[53] The growing numbers of smaller dwellings may be linked with a parallel shift towards greater specialisation in industrial work, linked in turn with the development of more capital-intensive technology.

The evidence of building and settlement therefore suggests that early mining was sustained along with small-scale agriculture but that this gave way in the early eighteenth century to far more specialised industrial work, generating in turn other by-employments that were themselves related to industry rather than to agriculture.

Development of settlement associated with the iron industry also exemplifies these functional patterns. Again, location of resources was a primary determinant in the location of the industry and its associated settlement. In Coalbrookdale, the blast furnaces required only a small labour force, and the iron-using trades were on no greater scale than in any small town. But both called for specialist work: from a very early date, the industry was highly capital intensive, both for the furnaces and the power system. The Hallen's Lower Forge works were making goods requiring very specialised skills – the making of frying-pans was the most highly skilled of all metal-working activity, and Coalbrookdale was only the third place in the country where this was done. A mid-eighteenth-century map of Coalbrookdale suggests that within the area of the furnaces and associated industrial buildings there were several buildings in which house and workshop were combined (fig. 6.6).[54]

But if some small-scale workshops were integral to the domestic units, other housing marks the separation of work and dwelling: terraces of small dwellings were characteristic of Coalbrookdale by the mid-eighteenth century, though these followed an earlier phase of isolated cottage-building in substantial plots.[55] Possibly the earliest terrace in the Gorge (Tea Kettle Row, 1735–42) was built for labourers in the ironworks.

Early industrialisation: land use and settlement

The development of industry in the countryside must be seen in relation to the whole economic structure of the area. Close connections between agricultural organisation and industrial activity have been demonstrated in other areas;[56] there is, for example, a clear link between the early enclosed landscape of pastoral farming and domestic industry. In some areas, industry developed as a supplement to poor agricultural livelihoods, and the early hosiery and textile industries, for example, appear to owe their origins to a pattern in which industrial work was taken up by an underemployed and poor agricultural population.[57]

In the Gorge settlement was associated with industry from the outset. Some industrial activity preceded the establishment of permanent settlement, and migration

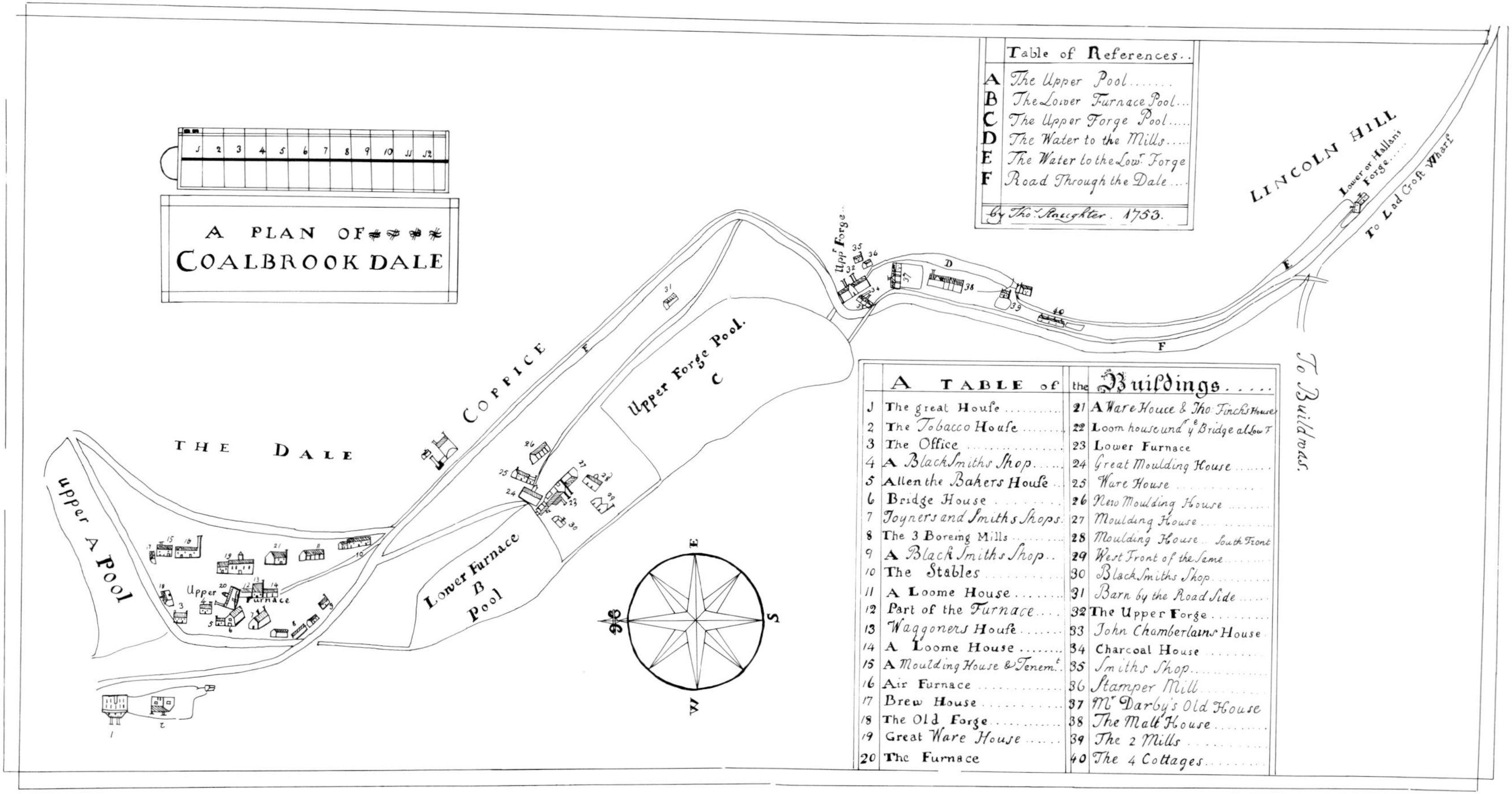

6.6 The close relationship of housing and manufacture is suggested in this plan of Coalbrookdale 1753 (IGMT)

into the area seems to have played a part in the early formation of a labour force. Those industries dependent on the availability of natural resources might be expected to have a different pattern of organisation and development from others which were more dependent either on agriculture or on craft.[58] In the Gorge, early industrial specialisation based on the vigorous exploitation of natural resources led to early prosperity, which was reflected in expanding settlement, the building of good-quality houses and the acquisition of more domestic goods during the eighteenth century. By this time, by-employments in the area were themselves industrial.[59] The classic model of proto-industrialisation, by contrast, supposes a land-hungry rural population drawn in to domestic, craft-based industries as outworkers for urban merchants.[60] Incomers to the Gorge had skills already, whether as miners or potters, and it was easy for them to settle in an area where land had low value, and where few legal constraints were raised against settlement.

But even some areas where industry was based on the exploitation of natural resources remained poor: the settlements in the Gorge are in marked contrast to those associated with mining in other areas of Shropshire such as Clee Hill – where the settlements remained isolated, the buildings less permanent, a dual economy more important. Mining technology remained relatively primitive and poorly organised.[61] In the Gorge, industries grew rapidly in scale and capitalisation and, above all, the economic basis of the settlements continued to diversify – not only was there an increasing degree of social stratification, but also increasing industrial diversification. By the late eighteenth century, a number of integrated industrial concerns had developed, and the communities had lost their original dependence on a single industry.

The importance of industrial specialisation in shaping the pattern of settlement can be paralleled in other areas. The North-West Leicestershire Coalfield had a similar history of development, from scattered settlements associated with bell-pit mining, to the creation of more concentrated settlements linked with more complex mining systems.[62] Similarly, in the West Midlands iron trade, the early development of the industry was characterised by a marked degree of specialisation and production for markets in pursuit of profit – and it was considered by contemporaries in the late seventeenth century to be a rich area. The Sheffield area shows a similar history.[63] Both were also characterised by the continued diversification of their industrial base. Specialisation and diversification are the common factors in the stability and prosperity of these early industrial communities.

The industrial landscape of the Gorge was marked by an early industrial specialisation associated with the settlement of new areas. Although these new settlements were organised for industry, they formed part of a wider pattern of estates and land uses. In the context of the estates as a whole, agriculture continued to be important, even benefiting from the expanding population on its margins. Nevertheless, there was a marked separation of the industrial communities from the primary settlements, and the new settlements formed a number of 'industrial villages' which continued to grow and diversify throughout the eighteenth century.

6.7 The beginnings of an urban landscape: Ironbridge and Coalbrookdale, *c.*1800. Map based on surviving and documented buildings (N. Smith)

Urban growth and change after 1779

The construction of the Iron Bridge in 1779 marks a turning-point in the settlement history of the Gorge, as the first stage in the creation of a small urban settlement, the town of Ironbridge. During the nineteenth century the Gorge developed as a coherent industrial district, and took on some of the characteristics of an urban area, with the new town of Ironbridge as its commercial and social centre (figs 6.7 and 6.8).

But the development of Ironbridge must be set in the context of wider changes in the economic and social landscape of the Gorge. As its industries diversified and increased in scale, so its social structure gained in complexity. These economic and social changes were directly reflected in the structure of settlement. From the late eighteenth century, the built-up area in the Gorge increased both in extent and in density as areas that had been settled early continued to grow. One new industrial village, Coalport, was established. At the same time, land was opened up for settlement of a new type. These changes were all part of the gradual urbanisation of the Gorge,

6.8 An urban landscape: Ironbridge and Coalbrookdale, *c.* 1850. Map based on the Madeley Tithe Map, 1847 (N. Smith)

which was most clearly expressed in the development of Ironbridge itself. This process entailed a radical reshaping of the industrial landscape.

Change in the early settlements

In the more complex industrial pattern of the late eighteenth century, the formerly close association of particular industries with particular settlements became less clear. New ironworks at Bedlam and Barnetts Leasow, for example, did not generate new housing in their immediate areas. The existing settlement pattern was capable of expansion under a vigorous housing market that was by this time well established. There was a continuous process of new building and development throughout the late eighteenth and early nineteenth centuries, but it was largely concentrated in the old settlement areas. The principal phase of expansion was at the turn of the century, characterised by an increase in the density of building on existing plots. This was the result both of new building and of the extension of existing buildings. Most of this development comprised small houses for workers.[64]

6.9 The south bank of the river in 1789. This sketch by Joseph Farington shows a series of settlements strung out along the south bank of the river. In the years that followed, this landscape came to be overshadowed by the rapid growth of the brick and tile industries (IGMT)

Broseley Wood provides an example of this increase in the density of settlement, and many of the surviving buildings date from this period: it became known as the poorest area of the parish of Broseley, said to contain its filthiest and most dilapidated houses and the subject of detailed scrutiny during the cholera scare of *c.*1830.[65] In

Coalbrookdale, a similar increase in the built area took place, partly through the development of housing by the Coalbrookdale Company, which signalled an expansion of output in the ironworks in the 1780s, but also through private development of small cottage rows, in a piecemeal process similar to the earlier pattern of development. Both areas had reached their maximum extent by *c.*1840.[66]

Similar developments took place in the riverside settlements, marking the shift from predominantly domestic to larger-scale factory production in the local industries. A sketch of the Coalford and Calcutts area in *c.*1780 shows extensive but open development (fig. 6.9).[67] But on the Tithe Map of 1840 the settled area was intensely crowded and densely built up. It was in this 60-year period that many of the larger and more land-hungry brick- and tile-works were established. They required both more labour and more land, displacing the older domestic economy and small-scale production. The small nucleated settlements gradually lost their individual coherence as the development of industry on a greater scale changed the pattern of land use and introduced new relationships between home and work. However, the location of new building was determined by the existence of earlier settlement nuclei: this was to lead to a loosening of the ties between particular industries and individual settlements.

As land uses competed in the areas of early settlement, the greater concentration of housing on pre-existing plots was an important aspect of the changing physical and social character of settlement. In Jackfield the subdivision of larger buildings, and the building of small terraced rows, were characteristic of the latter part of the century, as the increased size of the brick- and tile-works generated a demand both for more dwellings, and for houses which were not related to domestic production. Riverside inns were among those buildings subdivided into several dwellings. But the greatly increased scale of the late-nineteenth-century brick- and tile-works did not generate new housing on a concomitant scale – very few houses can be directly related either to Craven Dunnill or to Maws, suggesting that the growth of settlement of the eighteenth and early nineteenth centuries had provided an infrastructure of housing that was deemed sufficient.[68]

Over the same time, the social composition of the earlier settlements changed. The larger houses that had characterised the original areas were in many cases subdivided during the nineteenth century – The Tuckies, and the two timbered mansions on each side of the river were all subdivided into small units, reinforcing distinctions in social character which were brought about by patterns of new building. Lloyds Hall was a common lodging-house crowded with railway navvies and itinerant musicians in the census of 1861.[69]

In a parallel process, new land was opened up for housing of a rather different type. During the 1830s, settlement in Coalbrookdale grew according to a new pattern. Between 1827 and 1847, 8 houses were built on the previously open fields of Paradise, and a further 5 houses along the roadside – houses which were of a distinctly different type, and which were for the most part built singly in large plots.[70] At the same time, important changes were under way in the structure and organisation of the ironworks, with the introduction of new grades of skilled workers and an altered management hierarchy. The new houses were the homes of managers, draughtsmen and other members of the new middle class, which was one product of these changes.[71]

But while there were changes in the form and structure of the industrial settlements, until the development of Ironbridge the character of settlement in the Gorge remained that of a series of industrial villages. The significance of this is underlined by the development of Coalport, which was hailed as a new town, but which did not really differ from the early settlements in being largely dependent on a single industry, and in the scale, structure and pattern of its development.

The first settlement at Coalport was associated with the development of a transport interchange and communication network, focused on the Hay Inclined Plane of 1792 and the bridge of 1780.[72] The earliest houses were clustered around these two points (some 14 houses near the bridge), and there is little evidence for any earlier settlement in the area.[73] Very soon afterwards, the first china-works was founded, and further settlement followed, so that in 1802, Thomas Telford could write, 'The works at Coalport... have succeeded to a very considerable degree, and they are striking proof of the good effects of an improved inland navigation... houses to the number of thirty have been built and more still are wanted.'[74] However, rhetoric exceeded reality, little further development followed, and Coalport was only a small settlement in 1847, with some 38 houses associated with the chinaworks. It expanded little thereafter – only 9 houses were built later in the nineteenth century, and it was only the development of local-authority housing in the 1950s which added significantly to its size.[75]

Development was piecemeal, in small groups that were each built in several phases. In the second phase of development, plots were aligned with the road, but there is otherwise no suggestion of formal planning.[76]

Most of the housing at Coalport consisted of workers' cottages, though there were a number of more substantial dwellings (six large houses and an inn in 1847) associated either with management, or with small-scale family businesses (ropeworks, timber-yard, inn). Contrasting land use is suggested by the crowding of cottages onto small plots, and the layout of these larger houses, detached and in areas of garden.[77]

Coalport never became a town. There was little diversification of trades, very few shops and a small population. It remained largely dependent on a single industry.[78] Why Coalport remained so small may at least in part be explained by other changes which were taking place in the Gorge at about this time, and, in particular, the development of the town of Ironbridge.

The growth of Ironbridge

'The present town may be said to owe its origins to the construction of the Iron bridge, which here spans the Severn, and from which it derives its name.'[79] Although there is evidence for a substantial earlier settlement in the area, development that followed the building of the bridge did introduce some entirely new elements and involved a radical change not only in its immediate area, but also in the settlement geography of the entire Gorge. Although by the standards of other industrialising areas Ironbridge remained a very small place, the development of the Gorge in the nineteenth century acquired some of the characteristics of an urban area as functional differentiation came to be expressed in spatial structure. The settlement can therefore be compared with other urban industrial regions.

Chronology of development

The bridge was constructed in 1779, and shortly afterwards the Tontine Hotel was built and the market square laid out (fig. 6.10). The Trustees of the Bridge were closely involved in both these schemes.[80] A visitor in 1801 noted that 'Near the Tontine is a row of shops, in front of which a market is held weekly, on Fridays. The butter cross and shambles are neat slated buildings, supported by stone pillars.'[81]

6.10 Ironbridge, *c.*1835. The building of the Iron Bridge was the spur to a dramatic change in the organisation of the Gorge as Ironbridge grew up as a new town. In this lithograph, the Tontine Hotel, Market Square and buildings, and large new commercial and residential premises record the early stages of its urban development (IGMT 1980.304)

Several large blocks of shop buildings on the Wharfage and High Street frontages were established at the same time. In addition, a number of new cottages were built on small plots, often where buildings already stood. Behind the main street frontages, there were therefore areas that were intensely crowded with small cottages and workshops. This pattern of land use was analogous to the court development in some of the major cities, but was an inversion of that process, since the frontage buildings were actually new, and the crowded back building represented the intensification of existing development.[82]

A second phase of development began in the 1820s and was largely complete by 1847. This saw an increase in the density of the commercial area, with the building of new blocks of shops to form continuous street frontages in the area around the market square and the bridge. Much of this development was on previously vacant plots, though there is also evidence for the replacement of existing buildings.[83]

At the same time, a number of large villas and detached houses were built on formerly empty plots on the slopes of the Gorge. Development of land for larger housing continued into a third phase of development, beginning in the 1850s and continuing with intermittent building into the 1880s. Most of these early villas were built on vacant plots in areas which had formerly been associated with mining, and what appear to be blocked-up entrances to these mines still exist on Hodgebower and Church Hill. But in the latter part of the nineteenth century several such houses were built on the sites of rows of small cottages, suggesting a literal change in the valuation of land. Alongside the larger villas there were a number of more modest houses, built from the 1840s, also usually detached and in substantial plots, the houses of a prosperous lower middle class.[84]

The form of settlement

The pattern of land use in Ironbridge had been established by 1847. Its characteristic features were a densely built central area focused on the bridge and market square, progressively thinning out in density both alongside the river and up the slopes of the Gorge. There was a continuously developed street line on both sides of the bridge, and a degree of planned development, with the formation of large blocks, each containing a number of separate shop and dwelling units, but designed to be read as single buildings. The market square provided a commercial and aesthetic focus for this development, and was again a formal piece of planning, created by cutting back the natural slope to create a level platform. There were in addition a number of prominent public buildings, including a parish church and two schools.[85] Ironbridge was also a central point in the local transport network – a centrality further emphasised with the coming of the railway and the opening of a station in 1862.

All three of these elements (the pattern of development, the street lines, and the existence of a coherent centre) were entirely new elements in the settlement geography of the Gorge. Pockets of earlier development were absorbed into this new pattern, the new street frontages concealing a jumble of smaller buildings.[86] An earlier plot pattern survived as the tenurial basis for new development; these plots were big enough to accommodate a single villa with pleasure grounds, pairs of more modest houses, or clusters of small cottages.[87] There was no pressure for larger-scale development.

These physical patterns are closely related to the functions of the settlement. In the first place it was a commercial centre, with a market by 1801.[88] Development of this market and associated retail shops formed the first phase of development, preceding its establishment as a residential area. In 1837, Ironbridge was described as

> The merchantile part of the parish of Madeley, and here is the focus of professional and commercial pursuits. The weekly market, the post office, principal inns, drapery, grocery and ironmongery, watch making, cabinet making, timber and boat building establishments, the subscription library, subscription dispensary, branch bank, subscription banks, gentlemen of the legal and medical professions, ladies boarding school etc.[89]

This graphically conveys the extent and diversity of commercial and professional pursuits focused on Ironbridge – a range which is indicative of its urban function and suggests similarities with the constitution of other small market-towns. Almost all the buildings in the central area were in use as shops by 1847, and most appear to have been purpose-built.[90]

Warehouses, malthouses and inns comprised most of the development along the river to the west, dispersed along the river front and dependent on it. Though some of these (the Swan and the White Hart) predated the bridge, most were founded after it; all establishments were either built or rebuilt in the years after the bridge was erected.[91] These buildings also formed an area with a coherent character, dependent on river traffic and road trade, but with a lower density of land use than that of the central area, each complex requiring considerable working space.

The slopes of the Gorge were the principal location of new residential development, though clusters of cottages were crowded close to the commercial centre of the settlement, largely built around a nucleus of buildings that had been there before the bridge. On the slopes of the Gorge, as mines were disused, so housing developed. These slopes were marked by the lower density of occupation except on those areas where there was already settlement – clusters which stand out by their greater density, but the smaller size of the buildings. Many of the new villas were surrounded by large pleasure grounds and gardens.[92]

There was little other than light industry, or highly specialist production of consumer goods in Ironbridge itself, and different land uses were associated with different densities of development. The central commercial area was the most densely built-up, along with the nuclei of earlier settlement, while nineteenth-century residential development was characterised by its spaciousness. Although on a very small scale, Ironbridge therefore exemplifies some of the characteristics of an urban settlement, in terms of its spatial organisation and economic functions.[93] Its spatial arrangement contrasts with both earlier development and the other industrial villages of the Gorge. However, Ironbridge was only one component in the industrial landscape of the Gorge – this landscape can best be understood during the nineteenth century as an urban district, in which Ironbridge was the commercial and social centre.

Ironbridge and the Gorge

As a commercial centre, Ironbridge was marked by a concentration of retailing, including a number of specialist shops. The range of shops in Ironbridge and the density of their development contrasts with the small numbers of more general shops in the outlying settlements of the Gorge. In Ironbridge, the central area is marked by large purpose-built shops incorporating dwellings and storage space. Elsewhere, ordinary houses and house-types were adapted during the nineteenth century, or small additions made to existing buildings.[94] In the concentration of retailing, and the existence of a number of more specialised shops, Ironbridge is marked as a prosperous local centre. These were shops dependent on a central position and a wide hinterland – they were not the neighbourhood shops of working-class districts.[95]

Study of the distribution of retailing throughout the Gorge as recorded in successive trade directories confirms the central role of Ironbridge. In Pigot's Directory for 1846, 78 traders were listed under Ironbridge, 3 in Coalport, 11 in Coalbrookdale. That Ironbridge was indeed the commercial district for Madeley is underscored by the fact that Madeley itself listed only 23 traders, not many more than were recorded in Madeley Wood (19 traders). In 1851, three shopkeepers were listed in Jackfield in Bagshaw's Directory: in the same year, Ironbridge could boast bakers, booksellers, butchers, cabinet makers, druggists, hairdressers, hosiers and milliners among the range of its trades. The directories also show how these trades were concentrated in the central district of Ironbridge, with over half the shops listed located in Bridge Street, Market Street and the Wharfage, and all the specialist traders grouped in these streets.

But Ironbridge was not the only local urban centre fulfilling a significant commercial role: in 1846, 74 traders were listed for Broseley, which was also an administrative centre. Although Ironbridge was central to the Gorge, its hinterland remained small. It was one of a network of small towns in the coalfield, each of which became an important local centre, but none of which grew to regional pre-eminence.

Ironbridge also became the centre for a number of professional and public services. An infant school was built in 1831, and the National School in 1851, both in the central area. There were three chapels by 1835, and in that year a new Anglican church, Saint Luke's, was built by subscription and with a grant under the Church Building Act of 1818 – the provision of churches under the Act was largely confined to urban areas, again suggesting the new importance of Ironbridge as a small urban centre. It was designated as a separate parish in 1845.[96]

Some administrative functions were gradually relocated here as the formal and functional urban area was given recognition in the formation of a legal and administrative entity. Ironbridge had been recognised as a separate ward within the parish of Madeley since 1835. After 1889, it had its own district committee with the power to appoint subcommittees, and responsibilities in the field of housing and public health.[97] The police station and court room, built in 1862, were used by the committee, but in 1896 the Old Dispensary on the High Street was taken over and became known as the Municipal Buildings.[98] Public buildings enhanced the physical and social impact of the central area, while at the same time other distinctively urban services were introduced: the development of gas lighting provides an example.

A residential area with a character distinctly different from that of the previous housing grew up as part of this new town. The large numbers of villa-type residences on the slopes of the Gorge were the homes of a commercial and professional class. A number of houses in the area were the homes of industrialists who worked elsewhere in the Gorge. The Orchard was the most flamboyant of these, and was built by William Davies, whose brickworks were in Jackfield. By 1885, H.P. Dunnill of Craven Dunnill's tileworks was living there. Severn House was the home of Arthur Maw at about the same time.[99] John Randall described both the physical and the social character of Ironbridge in 1880:

> Clustering cottages are seen to perch themselves on ridges, or to nestle pleasantly in shady nooks half hid by rocks and knolls and trees . . . on points commanding panoramic sweeps of country, of winding dales and wooded hills, have sprung up villa-looking residences and verandahed cottages that tell of competence, retirement, and those calm sweet joys that fringe the eventide of prosperous life . . . from the Severn to the summit, the hill is dotted with villas, gothic and fanciful, fronted by grottoed gardens, flanked by castellated walls and orchards.[100]

By this time, therefore, Ironbridge could be viewed as a picturesque paradise, and house names and road names adopted during the nineteenth century reflect a pursuit of the picturesque (Belmont Road and Bellevue Road, The Orchard, Mount Pleasant and Prospect House were all first recorded in the late nineteenth century) (fig. 6.11). Residential segregation was never complete, partly because of the constraints imposed by previous settlement, but there was nevertheless a marked differentiation in the pattern of land development in different areas. By the end of the nineteenth century, Ironbridge was acquiring some of the popularity of a resort: the Crown Inn on Hodgebower, with its pleasure gardens, was a particular focus of interest.[101]

In the landscape of settlement, there was a growing contrast between the early developed settlements where small cottage building was crowded onto constrained plots, and new areas of development in which land was laid out more spaciously. This reflected an increased social stratification. Industrialisation entailed both an increase in the importance of services and the creation of more complex occupational hierarchies.[102] Ironbridge became a centre for commercial and professional services, and housing was developed largely for this prosperous middle class. Changes in the wider area of the Gorge helped reinforce this pattern. The early nineteenth century saw not only the reorganisation of the Coalbrookdale Ironworks, but also the establishment of larger industrial units – the brick- and tile-works of Jackfield in the latter part of the nineteenth century were far larger industrial complexes than their predecessors, and generated a new structure of industrial organisation.

So a distinct social area was formed within Ironbridge which was inhabited by people who worked not just in the immediate area, but also in other parts of the Gorge. Ironbridge developed away from directly industrial land uses, and came to be largely residential and commercial. Although housing for workers remained close to the sites of industry, other social classes tended to move away as new areas of land were opened up specifically for housing development. As a result, a new hierarchy of settlement was shaped across the Gorge.

6.11 Severn Terrace, Ironbridge, *c.*1900. 'From the Severn to the summit, the hill is dotted with villas, gothic and fanciful' (IGMT1982.2336)

Such a link between a differentiated social structure and the spatial organisation of settlement gives an indication of the extent of industrialisation and urban development and may be compared with the process historical geographers have described as Social Area Theory.[103]

One of the effects of this change was to create new relationships between the industrial settlements and the town of Ironbridge. The area which is now known as Madeley Wood, for example, which had been an important nucleus of settlement in the early eighteenth century, had become a working suburb of Ironbridge by the mid-nineteenth. Jackfield (fig. 6.12) was considered to be

> not all the world, but a very poor bit of the fag end of it . . . made up of old pit shafts, pit mounds, rubbish heaps, brick ends, broken drain and roof and paving tiles, dilapidated houses, sloughy lanes and miry roads

by H.P. Dunnill in 1870.[104] Residential segregation was therefore not just a consequence of the creation of a general housing market and a stratified society – new and larger industries not only used more land, but helped to create the desolation and decay described by Dunnill. This was particularly true of the brick- and tile-works of

6.12 Jackfield, *c.*1890. 'The fag end of the world'? (IGMT1982.2337)

the south bank. For those who could, there was every reason to leave the immediate site of industry.

This separation of different land uses and the changes in land valuation that it entailed modified the social geography of the Gorge as a whole, and the relationships between its various communities were reshaped. A new hierarchy of settlement and distribution of land use gave a coherence to the landscape of the Gorge.[105]

Ironbridge assumed the role of central place within the economic and social structure of the Gorge, largely on the strength of its commercial function. This commercial development owed a great deal to the building of the bridge and the road-building and improvements that were associated with it. Not only did they create a physical focus for the new town, but they also literally made Ironbridge geographically central to the Gorge. Trade directories make it clear that Ironbridge was the centre for a network of communication based on road and river by the early nineteenth century. One effect of this was to create a distinction between the centre and the periphery which again was a new aspect of the settlement pattern. This distinction was reinforced by the opening of the railway and Ironbridge station.

It is therefore possible to identify three major phases of settlement development in the Gorge in the period from *c.*1780. In the first place there was a considerable increase in the extent and density of the settled areas, which was followed by the emergence of a more diverse social structure, reflected in new house types and settlement in new areas. This process of partial segregation was enhanced by the development of Ironbridge as a small town centre, enabling the entire Gorge to attain a functional and social coherence as an urban industrial district.

The landscape reformed: into the twentieth century

At all times, the process of development had involved the demolition and replacement of existing buildings: few houses of the earliest phase of settlement survive; buildings were successively enlarged and improved, sometimes leaving little trace of the original structure; rows of cottages were pulled down and replaced by substantial villas in Ironbridge in the later nineteenth century; even earlier, the Darby family, in consolidating a private estate in Coalbrookdale, had been responsible for the demolition of several cottages in the area known as the Rope Walk Coppice.[106] Houses were absorbed into new factory complexes or lost to new transport systems: construction of the Severn Valley Railway, which opened in 1862, entailed a swathe of demolition in the settlements on the south bank of the river as it was pushed through the areas of densest development. The earlier tightly clustered nuclei were destroyed, and something of a linear pattern resulted, as the railway reoriented land use.

Powers available to local government were also used to modify the landscape in the interests of public health and land-use planning. The Public Health Act of 1848 made possible the establishment of Boards of Health and Inspectors of Nuisances, and additional bylaws.[107] As part of the Borough of Wenlock, and thus already part of an integrated local government area, there was little need or pressure for a Board of Health in the Gorge, but the first bylaws for the Madeley Sanitary Division of the Borough of Wenlock were passed in 1850. They were primarily concerned with the regulation of boundaries between public and private, with encroachments onto the highways through construction of cellars, sewers, vaults or drains; with the fall of roof-water to the pavement; with the regulation of nuisances, including wild dogs and obscene literature, kite-flying and football-playing; and the removal of waste. A system of inspection of houses and their sanitary provision was under way by 1851, laying the foundations for later concern with the state of the buildings themselves. By 1890, under powers conferred by the Housing of the Working Classes Act of 1885, houses deemed unfit for habitation could be closed, and lists of houses so designated were brought before the Sanitary Committee.[108] However, action did not necessarily follow inspection – no deep sewers were laid in Ironbridge until 1978.

Thus by the early twentieth century, local agencies of administration subjected the settlements to increased scrutiny following national legislation, and large-scale inquiries into the nature of the housing stock were carried out. Local records do not clearly set out the criteria under which housing was condemned, but a survey of housing in Ironbridge was carried out in 1881, and the detailed investigation of 72 houses that followed suggests that concern was focused on overcrowding and the adequacy of sanitary arrangements (in 1934 a slum was defined as a building which 'by reason of disrepair or sanitary defects . . . bad arrangement or narrowness of the streets [is] injurious to health').[109] Inevitably, the most densely built plots, which were also concentrations of the smallest dwellings, became the focus of attention. At first, action was taken on an *ad hoc* basis, listing individual houses as unfit for habitation, but a second stage of action saw the designation of small clearance areas. This process, finds a counterpart in early municipal action in other urban areas.

Although the sanitary inspectors were vigilant, it is never clear how effectively these powers were used locally: most housing clearance in the Gorge took place after 1950, and though long lists of houses deemed unfit for habitation were produced at committee meetings the evidence of maps suggests that most of the houses condemned survived until the mid-twentieth century.[110] Although there was some housing clearance in the 1920s, such direct action lagged way behind the system of inspections – in 1934, clearance areas in Church Hill, Severn Side, St Lukes Road, Jockey Bank, New Bridge Road and Wesley Road were designated but it was not until the 1950s that many buildings in these areas actually were demolished in tandem with the extension of a public housing programme.

The public housing programme initiated after enabling legislation was passed in 1919 modified the settlement geography of the Gorge both through new building and slum clearance associated with it. The first proposal for direct action in building was in 1920, and in November of that year it was proposed to build 24 houses to serve the Ironbridge and Coalbrookdale area. This scheme was dropped in the following June, and it was not until 1929 that any others were undertaken in the immediate area of the Gorge, though housing schemes nearby may have had some impact on the communities of the Gorge. The programme of public housing was drawn up for a wider area than the strict limits of the Gorge as land was made available for large developments in Madeley and Broseley – tenants of property in the Gorge were not necessarily rehoused in the immediate area.[111]

The housing developments in the Gorge itself were confined by the established structure of small plots, wherever land was available, but house-building and planning were carried out according to officially determined standards. Space was allocated to the new housing to provide a certain ratio of house to garden, and development was carried out according to coherent plans, involving a regular pattern of land use and layout. Access requirements oriented new housing to roads as far as possible.[112]

Some private building also took place, largely infilling in the existing plot pattern, or developing on derelict industrial land, more rarely replacing earlier housing: extensive bungalow building in Jackfield in the inter-war period colonised old spoil heaps, and in the Benthall Valley the site of the ironworks was laid out for development in small plots in which a number of bungalows and detached houses were built. The scale of these developments helped determine the adoption of planning standards of land use: individual buildings were inserted into existing plots, but areas of speculative building are identifiable by a new coherence in the organisation of space and orientation to roads. In the regulation of new development, as well as in the modification of older environments, the introduction of planning controls had a significant impact on the form of settlement.

Conclusion

The history of settlement in the Ironbridge Gorge can be traced from the first formation of small industrial hamlets, each with its own economic and social structure, to their eventual emergence as a small urban area. This pattern of change is interlinked with the course of development in local industries – early specialisation created the conditions for permanent settlement and investment in building; economic diversification, and the establishment of an industrial culture, had expression in the development of the Gorge as an urban area. The history of settlement belongs in this context of wider historical change.

Early settlement and prosperity expressed in building laid down a pattern for later growth and change. The components and organisation of the urban landscape were firmly influenced by earlier histories: the location of industries and of the earliest housing both formed a part in shaping an urban geography; the earliest settled areas became problems for twentieth-century planning. The twentieth-century landscape is still contained within the same pattern, and in spite of extensive demolition and clearance the earliest nuclei are still visible. These continuities form a major element in the present character of the area: how this inheritance was established and used in the past is the subject of the next chapter, which examines the processes of development which shaped the landscape of building.

7

SHAPING THE LANDSCAPE: PROCESSES OF DEVELOPMENT

Introduction

The industrialisation of the Gorge entailed major changes in the geography of land use and settlement patterns, outlined in the preceding chapter. This chapter traces the ways in which this geography was shaped through the operation of particular processes of land development and building. It shows that the various phases of geographical development were associated with particular types of land development, and that these in turn were linked with particular building types. The key to an understanding of the processes making this landscape lies in changes in the valuation and uses of land brought about by industrialisation. This chapter suggests that not only the buildings, but also the plots and their pattern of development, can be related to phases in the economy of land use.

Settlement in the first phase, up to the end of the seventeenth century, was characterised by the intensive use of isolated plots of land, the absence of control and regulation over the siting of development, and by a pattern of individual houses in small plots. There was a high replacement rate of buildings. The second phase, up to the late eighteenth century, was marked by both extensive and intensive use of land, the fixing of tenures and boundaries, development within existing plots as well as by extension of the built-up area. The survival of substantially built houses contributed to the emergence of a land and property market, and to changing land values. The third phase of settlement – between *c.*1820 and *c.*1880 – was characterised by the presence of an advanced housing market, large property-holding and tenant classes, new housing types which made different use of the plotted landscape, and the opening-up of new areas for building. In the twentieth century, public housing and speculative development were accommodated within the same land-holding structure but, as responses to a legacy of buildings, contributed to a revaluation of the housing stock.

Industrial settlement: shaping the plots

The early pattern of settlement, the landscape of the seventeenth and early eighteenth centuries, shows a pattern of dispersed development, of building in small, randomly distributed plots (fig. 7.1). This pattern can be traced in detail during the seventeenth

7.1 The Plott of Broseley, Samuel Parsons' map, *c.*1621 (redrawn, N. Smith)

century in the parish of Broseley, by means of a series of maps (figs 7.2, 7.3 and 7.4). These maps record a number of different types of plot. Some were defined by another land use, as, for instance, plots of land for building along a roadside, or close to an area of mineworking, or against the margins of marked-out fields or farms, but there are others which were carved out of open land or woodland and where the building of a dwelling preceded the marking of an enclosure. In still other areas the existence of small empty plots which were already held by many individual tenants suggests the early demarcation of areas for building. But in all these cases it is the presence of a house, or the intention to build one, which has created the characteristic small plots: the house is the nucleus of the plot.[1]

The overall form of the early plots was also variable – the size varied, as did the shape. Some were more or less regular rectangular shapes, others an irregular circular enclosure. Yet others were long narrow slangs alongside a boundary or road line (fig. 7.4). Although there are no maps for so early a period in other parts of the Gorge, study of the pattern of plots as it survived into the nineteenth century does suggest that similar variations occurred: in Benthall, for example, there is a series of small plots along the roadside, and an area of more irregular plot formation adjacent to an early mining area. In Madeley Wood, the pattern of plots associated with the earliest buildings shows some defined in relation to existing roads or boundaries, and others as islands in open land.

The shape and relationships between plots provide evidence for the organisation of previous land use, a haphazard pattern suggesting settlement on open ground, the more regular enclosures suggesting subdivision of existing marked-out fields. In Benthall and Broseley, for example, it is the plots that are taken from common or coppice land that are the least regular – those constrained by existing boundaries or features display a more ordered form (figs 7.2, 7.3 and 7.4).[2]

The definition of plots as enclosures for building brought about the subdivision of already existing larger plots or fields in many cases, though in others it began a process of dividing up previously open ground or woodland. By the early nineteenth century, the establishment of small units of land, the creation of a landscape of small plots, had gone beyond this immediate connection with building – by that time, the landscape of Madeley Wood was already divided into small plots before the hillsides were developed with the villas of Ironbridge.[3]

Most of the plots shown on the early maps contain a single building, showing that the first settlement took the form of individual houses dispersed in their own small plots (fig. 7.4). Other sources confirm this pattern: an engraving of upper Coalbrookdale in 1758 shows a scatter of small and isolated dwellings.[4] Maps made in the nineteenth century record marked discrepancies in the scale of development on different plots – even in densely built-up areas, there remained some solitary buildings.

Remnants of this pattern survive: many late-seventeenth-century or early-eighteenth-century buildings remain as isolated dwellings, and many of the terraces and groups of cottages were built gradually by extending an original single building. This pattern of building is the product of development by a large number of landholders

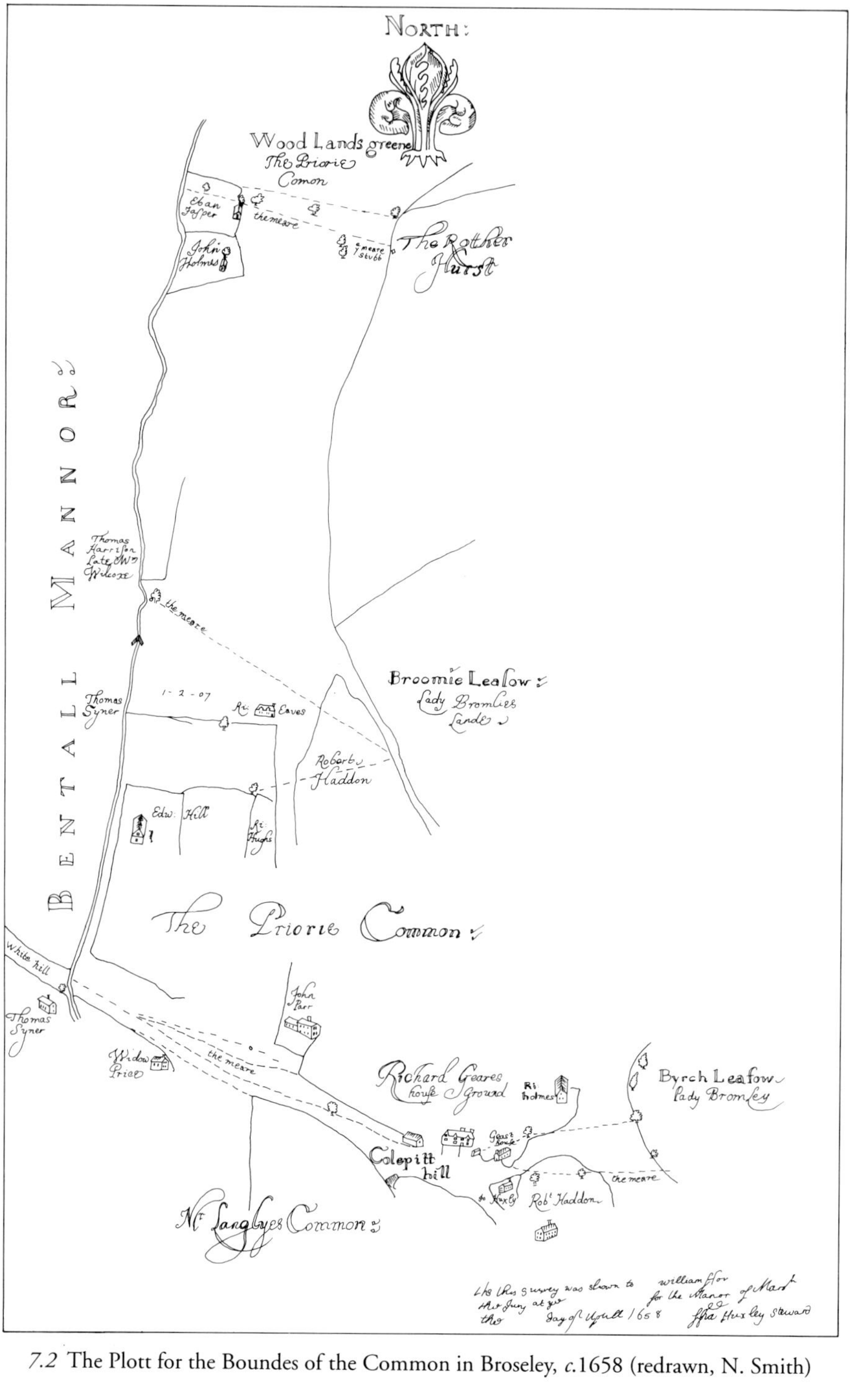

7.2 The Plott for the Boundes of the Common in Broseley, *c.*1658 (redrawn, N. Smith)

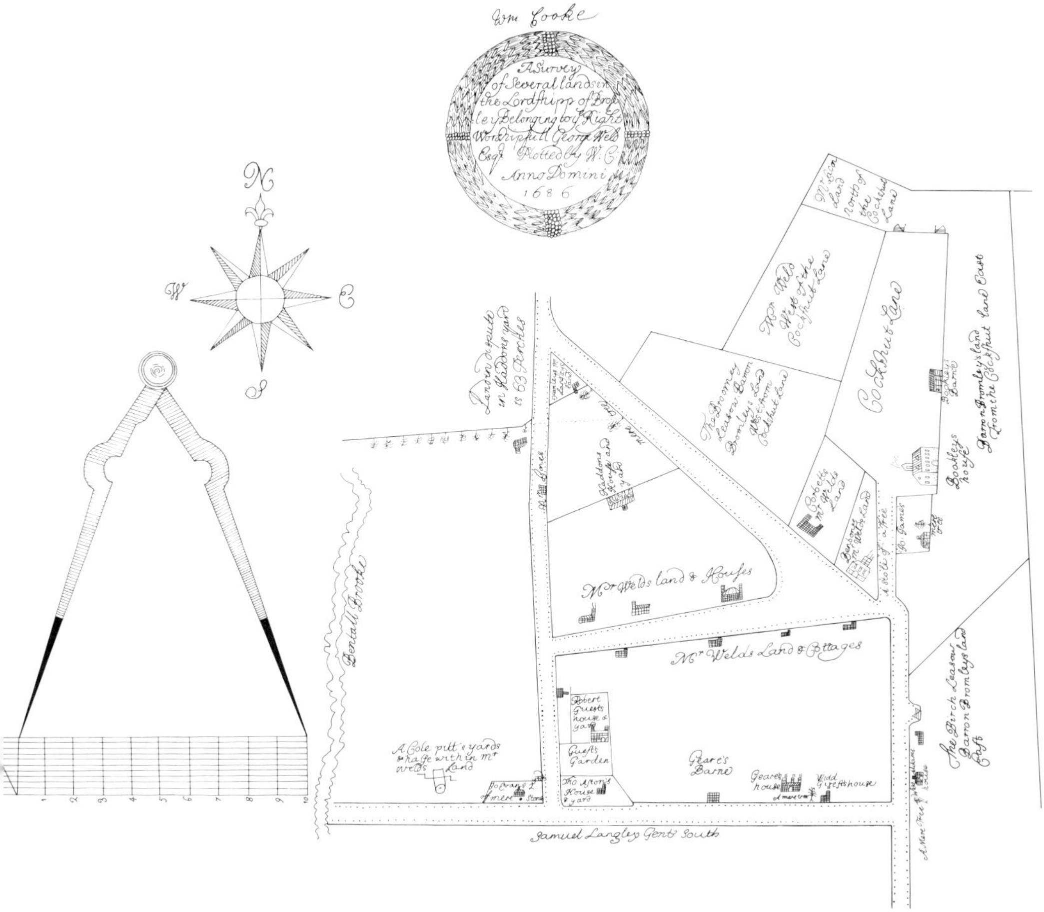

7.3 A Survey of Several Lands in the Lordshipp of Broseley, 1686 (redrawn, N. Smith)

(whether owners or tenants) building for themselves; the subsequent formation of terraces or pairs of cottages, by contrast, immediately suggests a different order of building organisation, more akin to speculation. That the earliest land development and building was carried out as a private process, with no evidence of organised housing provision, is confirmed by the documentary record.[5]

This piecemeal settlement pattern, of single buildings in small and irregular plots which had no coherent overall plan, is reminiscent of the landscape of squatting, defined as the development of land without tenurial rights. Squatting is associated with self-building – there is a tradition that refers to the house built hastily overnight.[6] The landscape of squatter settlement is characterised by random and dispersed development, the absence of clear relationships between the buildings, the often rudimentary nature of the building, and the frequency of alteration, similar in many respects to the landscape of the Gorge. Often, but not invariably, squatting was

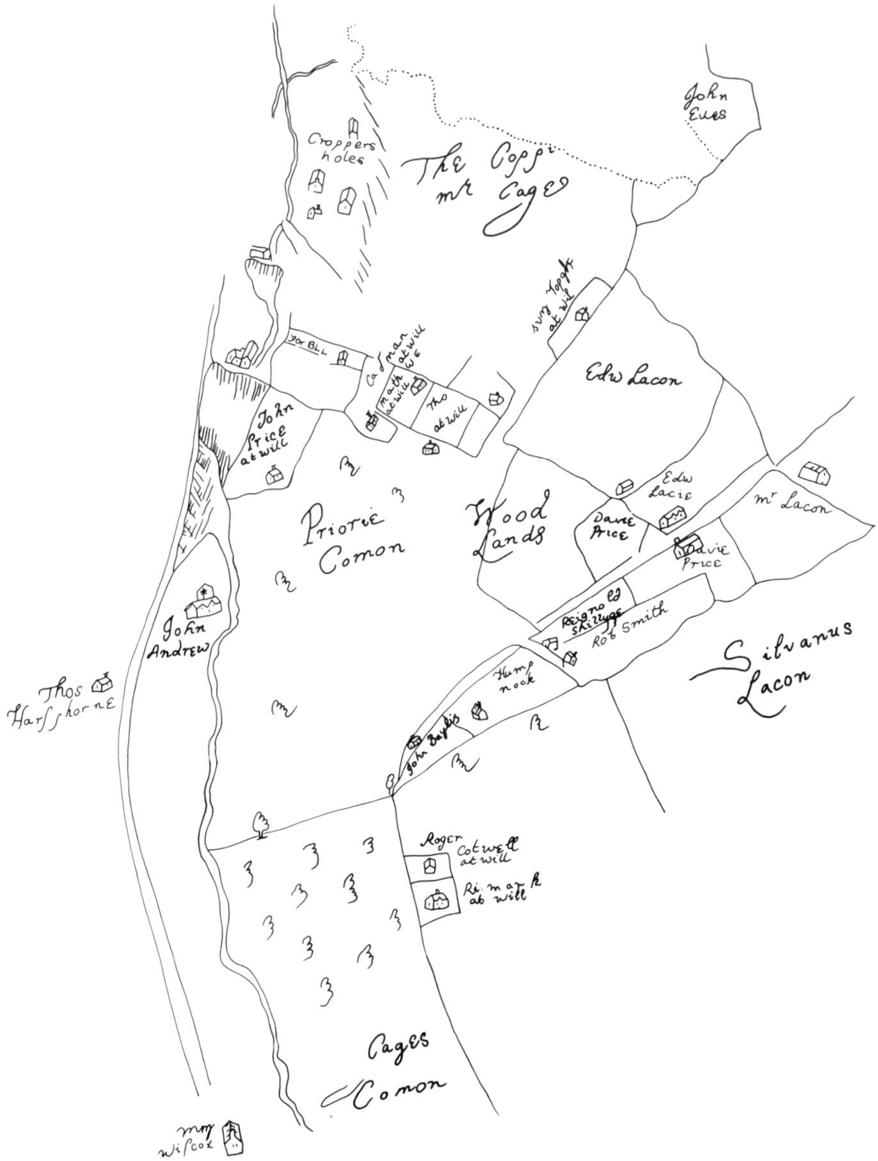

7.4(a)–(c) The plotted landscape: extracts from Samuel Parsons' map, showing the various relationships of houses to plots

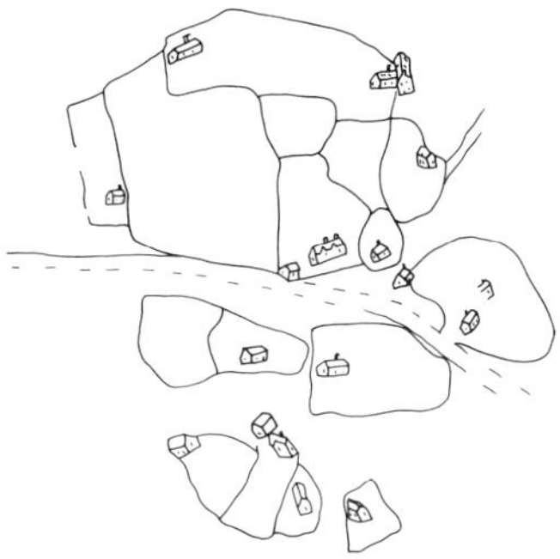

7.4(*b*)

7.4(*c*)

associated with that pattern of by-employment that frequently characterised the growth of rural industry.[7] The precise relationships between tenure and building, which can be established for parts of the Gorge, enable this model of settlement to be refined and modified, and show that the process of industrial settlement was as much the product of particular kinds of land use and valuation as of particular tenurial circumstances.

Some of the different tenurial arrangements under which early development in the various parts of the Gorge took place can be identified. The one that most resembles squatting is the settlement in Madeley Wood. There, development during the seventeenth century took place on the margins of the parish, in wood and common land, and the precise location of building may have been outside the control of the lord of the manor, in whose hands this land properly lay: there are a number of records which refer to unlicensed cottage-building during the seventeenth century. The existence of a large number of small plots which became freehold when the manor was sold after 1705 – and the very fact that the manor was sold off in small units of land – suggests that there were already a number of *de facto* holders of land in small units. On the other hand, it was the lord of the manor who was actively promoting the exploitation of mining in the area, so it was also in his interests that settlement took place.[8]

In Broseley early development took place within a framework of holdings created by the break-up of the manor. In this process, several estates and farms associated with the exploitation of mineral resources on the land were formed by the consolidation of smaller holdings. In the early seventeenth century, there were tenants on these estates holding several parcels of land developed with buildings and mineral workings. The commons had already fallen into private ownership – the early maps were made in order to define and clarify disputed boundaries between different holdings.[9] The details of tenure given on Parsons' map show that even as early as *c.*1620 there was a system of leasing and subleasing and some tenants held several houses. There are also references in contemporary documents which make it clear that some at least of the landholders encouraged the process of settlement, since it gave them a supply of labour with which to work their mines, and rights over mineral resources under common land.[10]

In Benthall, the parish comprised a single estate, with a landowner who was actively engaged in its exploitation: records from a dispute indicate that he also encouraged settlement on formerly common lands, so that, although the landowner did not himself engage in building work, and the new settlers were described by his opponents at law as squatters, this was squatting by encouragement.[11]

But even though tenurial rights were established over the land, the absence of coherent relationships between the plots and the buildings on them suggests that the detailed process of settlement was uncontrolled. The concept of the open village offers an alternative model for this unregulated, yet legally based settlement. This was noted first in connection with rural settlement, but has usefully been extended to industrial settlements by Barrie Trinder.[12]

An open village is one where there is no single landowner, where there may be several small freeholdings, and where manorial rights are fragmented. In this situation there would be no overall control over settlement, and building would be a vital source of income on freeholds too small for other viable uses. This model fits the parish of Madeley in the eighteenth century, since the manor was sold off in smallholdings after 1705, and the manorial rights were divided among several holders until reunited by Joseph Reynolds in 1781. Coalbrookdale had remained as a series of coherent farms since it had been separated from the manor in 1540, but in the early eighteenth century, one at least of these farms was sold off in small units of land for building, so creating a number of small freeholds.[13] On the south bank of the river, while Benthall remained throughout a single estate, Broseley comprised several estates and farms during the seventeenth and eighteenth centuries; and, although its landowners took an active interest in the exploitation of their lands, leasing and subleasing placed effective control over building in other hands.[14]

There was therefore no single tenurial structure unifying the different areas of the Gorge. The three processes of land development that can be traced by the early eighteenth century – encouragement by landowners, sales and leases, and *de facto* settlement through squatting – relate to this variety of tenure. The variations in the organisation of the plots of land, whether they were marked out for building first, or whether it was the presence of a building which defined a plot, may also have been in part the product of different tenures: but the resultant pattern was similar in each case – small islands of settlement, and buildings in irregularly shaped plots.

However, the absence of consistent variations in the settlement pattern, corresponding directly with tenurial differences, suggests that it was not ownership or tenure which determined in detail the form taken by settlement.[15] The irregular distribution and variations in plot form, and the absence of planned relationships between the plots, show that the precise allocation of plots for building was unregulated, even where the tenurial structure implied some sort of control. This pattern of development, which combines an established, though varied, tenurial structure with apparently haphazard development patterns, can be connected with the nature of land use associated with the onset of industrialisation.

The use of land in the Gorge area during the seventeenth century was intensive but not extensive – the development of land for mining and manufacturing left large amounts of underused or waste land which was not systematically exploited, and which could therefore be taken up for settlement. It was also associated with the creation of small blocks of landholding, where housing could become an important source of income. In a settlement oriented to farming, by contrast, land use may be extensive, and the availability of new land for building strictly limited. William Cobbett noted a similar distinction between farming countries in Kent, Hampshire and Sussex. In Kent, 'every inch of land is appropriated by the rich . . . a country divided into great farms . . . . The rabbit countries are the countries for labouring men. There the ground is not so valuable.'[16]

It was countries like these which might produce a pattern of squatter settlement, and of by-employments. In the Gorge, once land was taken up for settlement, this development became one factor changing the relative value of land: land with buildings was more valuable than waste. This helped to establish a market in land, and it was this, in turn, which helped to shape the second phase of settlement.

Building in a land market

The seventeenth-century landscape formed an uneven framework for later development. While the plots formed in the first stage of settlement clearly survived in some areas, in other areas there was a more radical change. Comparison of the series of seventeenth-century maps of part of Broseley with the Tithe map shows that in certain areas the entire early plot pattern was obliterated in later development, not by the creation of unified holdings permitting a more orderly pattern of settlement, but by another series of small plots (fig. 7.5). Early buildings did not always survive into the nineteenth century, and some may have been primitive structures that were quickly replaced. The short duration of the buildings themselves in certain areas may help to explain the absence of continuity of plot form: Hurst and Beresford have shown that in some medieval villages the customary rebuilding of houses once a generation was often associated with complete alignment changes if the buildings were of poor quality.[17]

Those late-seventeenth-century-buildings that survived into the twentieth century were nearly all built according to definable traditions. There is no suggestion in these coursed rubble and timbered houses of overnight or makeshift building. But these are the buildings which did survive, arguably the best of those built. There are enough of these to suggest that, by the late seventeenth century, there was money to be invested in building, and sufficient guarantee of stability to permit substantial investment in materials, skill and time. The extensive building phase of the early eighteenth century, producing buildings that were clearly built by professional craftsmen, represented a new phase in the development of settlement and in the formation of the landscape. These buildings constitute an important capital investment.[18] The ability to build therefore depended on a degree of prosperity, and the shift to permanent professional

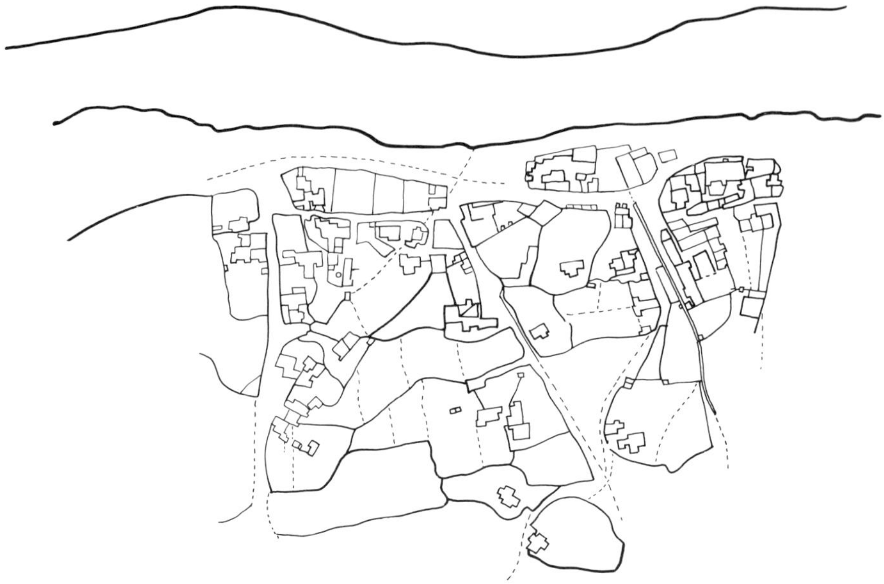

7.5 Extract from the Broseley Tithe Map, 1840. The plot pattern continued to shape land use during the nineteenth century. At Lloyds Head, plots with only a single building contrast with others crowded with development

building was dependent on the accumulation of capital resources, which itself was a product of the economic development of the area.

The construction of permanent buildings helped to determine the survival of the plotted landscape, and the level of investment in buildings was an important factor influencing continuity of landscape features. The significance of the investment represented by buildings is implicit in their survival, but it is also evidenced in the adaptation of buildings for new uses: throughout the Gorge, early buildings have survived through adaptation, by subdivision or amalgamation. In Coalbrookdale, for example, a seventeenth-century building, once a 'steelhouse', was in use as a malthouse in 1786 and had become dwellings by 1805.[19]

The seventeenth-century pattern of plots and settlement survived better in certain areas than in others, and permanent professional building was established at different times in different areas. These variations express differential investment in building, which can be attributed to highly local economic circumstances, and to differences in tenurial organisation.

Investment was influenced by the nature of tenure: studies in other areas have shown that rebuilding can often be linked with security of tenure. An important factor was the possession of an interest in property 'sibi et heredibus', and fieldwork elsewhere has suggested a close correlation between manorial policy on tenure and housing types and conditions.[20] In the Gorge, the small plots on which building took place were often subordinate to the principal tenurial divisions and were not always secure holdings: before the sale of the Manor, cottage building in Madeley Wood was unlicensed, while in Broseley Wood tenancies at will characterised some areas but not others during the seventeenth century. These were the smallest and least secure units of holding, and it is on these that there is most evidence for substantial modification to the pattern of plots in later development.[21] Concomitantly, the survival of buildings in greater numbers from the early eighteenth century may be evidence for the introduction of securer tenures: in Madeley Wood, for example, it may be no accident that the creation of small freeholds in the early eighteenth century was immediately followed by a rapid phase of new building.

At the same time, house-building depended on the availability of capital resources: late-seventeenth- and early-eighteenth-century buildings represent a substantial investment. The time-lag between the onset of mining and manufacturing and the persistent survival of housing may be an expression of the time taken to accumulate sufficient resources to invest in professional building.[22]

Development from the late seventeenth century was therefore the result of renewed investment in building, relating both to tenurial and economic circumstance. Not only were houses built substantially enough to survive, but settlement was also continually expanding. Comparison of the three seventeenth-century maps of Broseley shows something of how this expansion came about: open spaces between the first islands of settlement had also been enclosed for building, and property divisions formed recognisable roads, where the beginnings of a regular building line was established.

Similar processes can be traced elsewhere as more plots were created, and development on existing plots intensified (fig. 7.6). There was a concentration of building in certain areas, property boundaries formed footpaths and were sometimes coherently aligned as lanes; new land was also occupied by development, though there were still large areas which were not taken up with building, and pockets of empty land that survived between settled areas.[23] By the early nineteenth century, much of the land in the vicinity of the settled areas was also divided into small fields and plots, showing a definition of tenure and land use that embraced the whole landscape.[24]

If the siting of the first building plots was unregulated, the pattern of building that emerged during the century that followed, characteristically of dense building on small plots, suggests that this was very quickly formalised. The crowding together of buildings from the early eighteenth century was the product of constraints on the availability of new land for settlement, as the limits of tenure were more clearly delineated, through ownership or through conflicting uses. The careful definition of boundaries in title deeds from the beginning of the eighteenth century is evidence of the concern to establish and protect the limits of tenure. In most areas there is a clear boundary between areas used for building and areas that remained as fields or were used in the context of industry. This shift to an intensive land use marks a second stage in the development of industrial settlement, and it distinguishes the Gorge from other squatter settlements such as Clee Hill and Snailbeach in the south of Shropshire where precise uses and tenurial responsibilities were never firmly delineated.[25]

The development of intensive use of land on an extensive scale was a product of

7.6 Jockey Bank, Ironbridge, 1976. Early buildings served as the nucleus for piecemeal development as settlement rapidly expanded, crowded in on plots that were already occupied. This terrace was formed in such a way, from different buildings constructed at different times (*Shropshire Star*: IGMT 1982.2189)

the rapid industrialisation of the area during the early eighteenth century. As the scale and physical extent of land use in industry extended, so were changes in the value of land brought about: the subdivision and sale of the various estates in the Gorge during the seventeenth and eighteenth centuries is itself an indication of the importance of land as a commodity with a definite value. The use of land for industry also created a direct pressure on space available for building – in Coalbrookdale, for example, it was the steep slopes of the valley which could not be used in an industry dependent on water-power where housing concentrated. A hierarchy of investment, in which there was a calculation of profitable land use, may have helped to shape the settled landscape.

The growth of industry, and a population largely employed in mining and manufacture within the Gorge, encouraged the consolidation of agriculture in the adjacent countryside, and accentuated a division of land use. The industrial settlements represented a large local market with some quite specialised needs – horses, for example. The survival of productive farmland may account for the slow release of land in lower Coalbrookdale for building, an area not developed before the early nineteenth century; in Benthall, clear boundaries between agricultural land, industry and housing were maintained throughout the nineteenth century. The industrial development at Coalport and Blists Hill, which did take place on former agricultural land, is quite distinct from the pattern of development elsewhere in the Gorge.

At the same time, a growing population dependent on industrial work rather than on agriculture created a market for small houses which could be built on smaller pieces of land as work took place largely outside the home. The subdivision of plots to accommodate new building became possible: a similar process marked the evolution of settlement in the textile areas of Lancashire, and has also been traced in mining settlements in Cornwall.[26]

A particular pattern of development emerged as a response to this demand, constrained by the organisation of land use and holdings. This pattern can be reconstructed through studying the relationship of the buildings of this phase to a plot structure. In some areas, there is a close link between plot shape, building layout and ownership – buildings and ownership both respect the plot boundaries – suggesting that the plot was at once a unit of tenure and of development.[27] But even within a small area there could be a marked discrepancy between different plots – one plot may have had only a single building, others have been very densely developed.

There was, in addition, considerable variation between plots as to how land was held. Some had a single owner and a large number of tenants; elsewhere, coherent units of land were divided among a number of owners (or primary tenants in areas of subtenanting), suggesting that an original plot had been sold off in small units, either as development took place or in advance of it. The pattern of holdings in the Ironbridge area in the mid-nineteenth century, for example, also shows that some of these holdings were achieved as the result of amalgamation of small plots: they show a discrepancy between the way the plots were built up, and plots as units of tenure. The plots had been developed in smaller units and subsequently amalgamated as units of investment.

The transactions represented by the amalgamation of holdings and their fragmentation suggest the importance of a property market. Variations in the size of the units of tenure may reflect relative values of different areas of land: in the Madeley Wood area, the smallest plots and units of tenure were in the centre of Ironbridge, which was both an old-established area, and one where the value of land increased rapidly following the development of the town. Elsewhere, a highly fragmented plot pattern was the product of the age of settlement – in the earliest developed areas successive divisions for inheritance and sale had produced very small holdings.[28]

The earliest pattern of a single house on each plot was modified by extensive building but the variation within areas suggests that this was at least in part an individual investment decision, though the economic development of different areas of the Gorge also determined land values. Housing plots were the smallest economically viable units of land, and an investor had a notional choice between a single building commanding a high rent (as in the commercial premises in the centre of Ironbridge, where the individual units of holding were also very small, and the larger detached houses) and a large number of small cottages.

The expansion of settlement during the eighteenth century was shaped by the overall economic development of land within the Gorge, reflected in the more intensive exploitation of the plots. The specific responses of the inhabitants of the

Gorge can also be charted through wills and title deeds. Among the processes responsible for shaping this intensification of settlement, the development of family property was pre-eminent. Limitations on existing house type were soon felt as families grew, and first houses were subdivided for children, or additions made, representing above all an expansion of economic opportunity.[29]

David Levine and others have shown that the demographic regime typical of proto-industrialisation was one of population growth through natural increase reflecting a falling age of marriage; this was itself a result of a decreased dependence on land for a living, and the opportunity for good livings based on wage labour outside the home. Development in the Gorge fits this model: M.D.G. Wanklyn has shown that the substantial increase in population during the later seventeenth century was largely owed to natural increase.[30] Wills, probate records and title deeds stress the importance of family formation in the creation of new households, employment in mining and manufacture being conducive to early marriage. These new households were a mirror of individual prosperity, and necessitated an increase in building.

It was the ability to provide for a second generation that underlay much of the early development; this was often concentrated within the original plot, often involving changes to the original building. Wills and title deeds show how new households were created through subdivision or through additions to existing dwellings. A will of 1728 arranged for the house to be divided between wife and son, another of 1746 between daughter and grandson; in William Botteley's will of 1762, the kitchen, cellar and parlour already in his occupation were left to a son, while the other part of the dwelling-house, known as the little room, was left to another son. John Bayley, collier, in a will of 1712, detailed the arrangements for the inheritance of the 'tenement in which I live and lately purchased of assignees of Comerford Brooke', which had been divided into three separate dwellings: his wife was to get the part he already lived in, being the chief dwelling, his son Jonathon to get the little room with loft over, his son Edward the middle part where he was already living, and his daughter Mary the brewhouse and chamber over.[31]

The ownership or lease of several houses and holdings was also common by the early eighteenth century. Descriptions in wills indicate that sometimes separate houses are clearly adjacent, and additional buildings had been built within the same plot: the will of John Bayley in 1784 refers to 'two houses, one I dwell in, the other adjacent occupied by my son in law'. Richard Cox, collier, 1774, had a freehold house and another adjacent house which he had bought from his uncle, and a cottage he dwelt in. Other wills suggest that sometimes quite detached plots of land were held in single ownership: in 1782, for example, the will of George Hodgkiss left the rents of freehold and leasehold houses at Dace's Hill and the Brockholes, and a house in the Lloyds.

These sources show that in many cases the occupier was also the builder, and there are numerous references in wills to houses built or caused to be built: 'the house I have repaired and the new building added to', Edward Boden, collier, n.d.; 'I have erected a house and other dwellings', William Botteley, 1762; 'The above 5 premises near Ironbridge have been built or set out by me', Michael Hodgkiss, 1796.[32] Title deeds also indicate that building was the responsibility of the tenant; there is some evidence of rationalisation after the event; for instance, a lease of 1727 for land in Broseley Wood shows that the tenant, Francis Eaves, had already built a house on part of the premises. Other leases were made for the purposes of building, or refer to enclosures from the waste. The property holders were colliers, watermen, barge-owners, smiths, furnacemen. They almost certainly did not literally build their own houses, which in their design and construction more often suggest that they were the work of professional builders (see chapter 8).[33]

These records show both a process of development within a single plot and the holding of several plots. They show that, development within a plot could come about either through subdivision or additions to an existing building, or through the construction of a separate dwelling. They show that, while direct building to house new generations within a family could be important, property could also form a simple financial investment, through leasing. The will of Michael Hodgkiss in 1796 clearly states this in referring to five leasehold houses which he had built or set out for the benefit of his children. Subdivision of holdings within a family eventually merged with the use of property as a form of investment, and property left in wills to children may often have been tenanted. At the same time, migration to the area created a need for new building, and new opportunities for the use of property as income.[34]

Documentation for the sale and development of Furnace Bank Farm, Coalbrookdale, during the eighteenth century also shows the value of property as investment, and the growing importance of a property market. Unlike the pattern of settlement elsewhere in the Gorge, this farm was progressively sold off for development in marked-out and defined plots, though in common with the rest of the Gorge there was no systematic relationship between the plots so defined.[35] All these plots were described as 'marked out and set separate' – they were almost certainly sold for the express purpose of building, and were taken, not from farmland, but from coppice woodland.

Furnace Bank Farm was acquired by a yeoman, George Dukeshill, in 1704 at the break-up of the manor of Madeley, and he left it to his son, a carpenter, in 1717. It comprised a farmhouse, a cottage, 42 acres and 14 acres of coppice. The first recorded sale was in 1715, a small piece of land (60 yards by 60 yards) which was sold to Abraham Darby for £24. Other sales soon followed: in 1717, 2 acres were sold to John Thomas, and in 1720 George Cartwright had acquired a piece of land in Brick Kiln Meadow for 50 sh. In 1738 the manager of the ironworks, Richard Ford, acquired a large plot on which he built Rosehill House, Tea Kettle Row, and the summer house called Mount Pleasant. In 1738 and 1743, Abraham Darby bought two small plots which were used for the development of Sunnyside House. In 1748, George Cranage bought a small plot and Thomas Cranage and John Thorpe also bought small plots in 1741 and 1742.

Subsequent development on these early plots can also be traced. Abraham Darby sold a small piece of the Sunnyside estate to George Cranage in 1748 and this remained the property of the Cranage family until 1809. It was divided between George and his son Edward in 1772, and the stable was converted into a new dwelling.

Land acquired by Andrew Cartwright in 1720 was also developed as a family property: in 1745 the property was mortgaged, described as 'lately extended to form 2 dwellings'.

In 1741 Thomas Cranage bought a piece of land which had four tenanted houses on it by 1763. These properties were later sold off individually. John Thomas bought a plot in 1717 which remained as family property, divided among children and developed in consequence with at least two houses. Outbuildings were converted to house at least three other households by 1794, and one house had been converted into four tenements by 1810. John Thorpe's single plot of 1742 contained five houses by 1788 and nine by 1821.[36]

The role played by private building on such small units of land in the development of the Gorge links it with squatter settlements or open villages. But what distinguishes the pattern of development in the Gorge from that of squatter settlements is both the extent of development and the relative formality with which alterations and additions to buildings were made. Building may have been piecemeal, but the final groupings of buildings more often form regular terraces or pairs than the random clusters of buildings with confused property divisions that are found in other areas of squatter settlement.[37] In the Gorge, the use of professional builders, suggested by the coherence of housing design and construction after the early eighteenth century, is an index of the prosperity of the area. It is this that marks out the Gorge as a prosperous industrial region in the eighteenth century.

The transition from development based largely on a family economy to one reflecting a more general housing market can also be traced in the buildings of this period: there is ample evidence of formerly single cottages subdivided to form two dwellings; there are pairs of cottages or small terraced rows which have clearly been built piecemeal, by additions to an original dwelling unit; there are houses where a front or rear wing has at some time formed a separate dwelling; there are plots containing small clusters of building as additional houses were erected on the same piece of land.[38]

This pattern of building, in which a single house became the nucleus for additions and modification, reflected the role of the occupier as developer. But the growing importance of a market in property, shown in the increasing numbers of transactions recorded in title deeds and wills, is reflected more fully in changes in the organisation of building: from the early eighteenth century, pairs or terraces of cottages were also being built as coherent, designed developments, early speculations in building (fig. 7.7). By the end of the eighteenth century, most new cottage-building took the form of deliberately designed pairs or rows of cottages – and small individual houses were relatively rare.[39]

Like the holdings built up piecemeal, these developments were also confined to single small plots, and a row of as many as six cottages was rare. But planned development of this sort represents an organisation of resources quite different from gradual additions to a single house: in the one row which is fully documented, it is clear that the cottages were built as a speculation, all leased out from the outset. The distinction between those houses which were built as an individual development and

7.7 Madeley Wood, 1952. Single dwellings, buildings extended gradually, deliberately designed pairs and terraces, all contributed to the makeup of settlement in the Gorge (Denis Bird: IGMT 1982.246)

those which were part of a small planned development was to become an important element in the social hierarchy of housing types.[40]

Initially it was the first occupiers of the land who played a major part in its development, and it was a system of owner-occupancy or secure tenancies which underlay the early growth of settlement. But from an early stage property was also held as income, though the boundaries between family property and property as income and investment were not always clearly delineated. As early as 1705, speculation in property was recorded in Madeley Wood,[41] and the use of property as a form of income quickly meant that there was an ever-growing class of tenants. In addition, the survival, adaptation and reuse of earlier buildings also suggest that the earliest connection between occupancy and building work would have been quickly modified. The development of a property market represents a very substantial change in the organisation of building, as property came to have a value independent of its immediate use.

The pattern of development in the Gorge up to the end of the eighteenth century suggests that it is changes in the organisation of land use, and the concomitant changes in its value, which strongly influenced the ways in which settlement developed. If there was an early stage of primitive building on ill-defined plots of land, akin to squatting, this was quickly superseded by increased investment in building, by rapidly developing tenure and land use and the development of a market in land and building. It was the development of mining and manufacturing during the late seventeenth and eighteenth centuries that created the conditions for the growth of settlement, in providing good livings, bringing about changes in the value of land, and in shaping the organisation of land use into which settlement fitted. Development of housing itself also helped bring about a reorganisation of land use and its value – buildings that became permanent by virtue of the size of investment they represented were an important element in shaping the landscape of industry.

Formal development: Coalbrookdale and Coalport

The origin of the idea of aesthetic design and planning for new settlements has been assigned to the eighteenth century:[42] an important contribution to the emergence of new model settlements was the creation of planned industrial communities, where development was undertaken by the major landowner or industrialist.[43] The housing provided in Coalbrookdale by the Coalbrookdale Company has sometimes been taken as an example of this, the creation of an industrial village by paternalist ironmasters.[44] However, company-house building took place only on a very small scale, and did not initiate a new settlement type: in the 1780s, four terraces were built, comprising 30 dwellings. The company did not repeat this direct intervention in development again. The Coalbrookdale Company was never a major owner of land for housing, and until the late nineteenth century concentration of ownership was restricted to those areas covered by the works themselves. By the 1780s, Coalbrookdale was already a sizeable industrial settlement, and its earlier development had included a number of small, speculatively built terraces. The scattered location of the four company terraces, which did not aspire to a formal development plan, suggests that the company was constrained in its acquisition of land for housing.[45]

The short phase of direct building was characterised by the use of a coherent building type, with a form which was repeated in each of the four terraces built. It is this coherence that shows that the building of these terraces was a direct intervention by the company, quite distinct from the development of privately held plots. The records that accompany the development show that it was undertaken as a form of investment: ten years after they were built, at least one of these terraces (Carpenters' Row) was sold at a time when the company was looking for ways of reducing its debts.[46]

Prior to their involvement in building at Coalbrookdale, the company had already engaged in building elsewhere. Both at Newdale and at Horsehay, symmetrically designed rows of houses had been built in the 1750s and 1760s. The rows at Newdale adopted a back-to-back plan, and were conceived as part of the overall design of the ironworks complex. At Horsehay, the cottage rows were characterised by a unity of plan and form, but were in no sense a planned village. Horsehay also embraced a more haphazard squatter settlement on the neighbouring Lawley Common. In each case, building development by the company was carried out on a very small scale, and, while individual buildings were carefully planned, there was no attempt to create an entirely planned layout.[47]

Coalport, the deliberate creation of William Reynolds, was a rather more ambitious project which has sometimes been described as a new town: 'The whole of this lively and beautiful place with its erections, belongs to William Reynolds, who has still continued planning and erecting for himself and others, sparing neither labour or expense to render it as complete as possible.'[48] The infrastructure of settlement was laid down by Reynolds, based on a transport interchange and a new river crossing, and investment in manufacturing led to the development of the Coalport China Works. As a settlement, however, Coalport grew very little, and remained at best a village. Moreover, the process of development was not subject to an overall plan: Reynolds was never the sole landowner, and building plots were not formally demarcated. A first phase of development saw the clustering of buildings together on small plots; a second stage of development made reference to a road line which followed a property boundary, but the buildings themselves were not coherently planned or aligned in relation to it.[49]

The leases suggest that in some cases building took place in advance of a specific agreement: one referred to a house and kiln built 'in expectation of a lease'. Reynolds' first lease and sale gave authority to build a range of building types, while tenants or subtenants were free to build on allotted land, subject only to the need to maintain existing and subsequent building in a good state of repair.[50] Building rows that survive were built in several phases and without use of a coherent design, in contrast to the development of company housing in Coalbrookdale. Only one row, a terrace of 14 houses, had any suggestion of a coherent design – and even this was built in phases.[51] The organisation of building was not subject to overall control, and was carried out by a number of small property holders under lease.

Development at Coalbrookdale and Coalport may be compared with other areas where housing was provided by the major landowner or industrialist. House-building was an expedient to attract workers, but was often undertaken with reluctance.[52] Most estate settlements are attached to a single industry, but even many of these demonstrate a reluctance to invest in extensive housing schemes except where absolutely necessary. Under-capitalisation of industry left nothing to spare for philanthropic provision.[53] Unlike early factory colonies in the textile industry, or settlements associated with the iron industry in Wales, the Gorge had a far more diverse economic base: throughout the Gorge, the survival of a highly fragmented tenurial pattern, the absence of any large-scale planned development and the large numbers of small building projects suggest the wide distribution of resources sufficient to carry out building on a limited scale.[54]

Both these company-inspired building projects make best sense in the context of changes in the organisation of building work which were happening anyway, as small-scale speculation by a distinct property-holding class superseded more individually focused building work. Both the Coalbrookdale Company and Reynolds at Coalport acted as speculative developers, on a scale slightly larger than the normal practice, but without making a major modification to the overall shape or type of settlement.[55]

An inheritance of land and building: development during the nineteenth century

By the standards of other industrial and urban areas, development in the Gorge after *c.*1820 took place on a very small scale – absolute numbers of new buildings were low, there was no large-scale planned expansion, and only limited speculative building. But, in spite of these differences in scale, development during this period did shape an urban landscape, which made use of an established pattern both of landholding and of building.

As in other urban areas, the pattern of land-ownership was vital in shaping the form of subsequent land use. In other towns, early urban building on fields and gardens fitted existing blocks of landholding, and suburban growth was also constrained by the existing field pattern: it was only where the units of ownership were big enough that any semblance of overall area planning became a possibility.[56]

Development during the late eighteenth century and early nineteenth century frequently entailed the subdivision of existing blocks of land and the creation of many small freehold or leasehold plots, reflecting the formation of a market in land as its value changed. Even where initial plots were themselves small, as in urban gardens, for example, building was often followed by sales, creating further subdivision.[57] Where development was part of a general building boom and rapid expansion of settlement, it was informed by a notion of planning, and building was carried out within a coherent framework.

In the Gorge, by contrast, the same shift in value and fragmentation of holdings had taken place in an earlier period, when the small scale of development, the diverse economic base, and the absence of a pre-existing urban framework had combined to create a settlement formation which comprised many piecemeal units of land, divided into small holdings with many owners. In Ironbridge, for example, there were large numbers of property owners with an average of three houses each at mid-century.

This is superficially similar to the pattern found in other areas of the country, though elsewhere it is often the result of sales of more coherent blocks of land as they were exploited for building.[58] But in the Gorge the subdivision of holdings reflects as much the age of a built-up area as the method of its settlement. Development in the nineteenth century did not often involve further subdivision of already small units of holding – equally, there was very little amalgamation of holdings. This was one factor ensuring a degree of continuity in the structure of the landscape, as the small scale of ownership discouraged its fundamental reorganisation in new development.

So, although the presence of many small units of landholding was a characteristic common to many urban industrial areas, the timing of the formation of that pattern was crucial in the differential development processes of the Gorge and towns like Nottingham or Leeds, for example.[59] In these other towns, divided ownership was the result of subdivision of already planned layouts – in the Gorge, these divisions had been formed without planning. The contrast is underscored by the fact that in other urban areas roads and building plots were commonly set out in advance of building, whereas in the Gorge this did not happen.[60] The turnpike roads of the early nineteenth century – Madeley Bank, Jigger's Bank, and the Ladywood Road to Broseley – cut across existing plot boundaries but were not systematically developed with new building plots. While some new building in Ironbridge aligned itself with the turnpike, more often the established pattern of plots remained the basis for building work.[61]

Even the new and more self-consciously urban development of Ironbridge was marked by the use of an existing pattern of plots and landholding. This process differed from the earliest settlement formation in the Gorge, since when it had been house-building itself which had entailed the creation of a landscape of small plots, while in the nineteenth century the plots preceded development with housing. By the mid-nineteenth century, the landscape of Madeley Wood was almost completely apportioned in small plots; waste land was scarce, confined for the most part to the margins of the settled area. But the pattern of plots was not itself uniform: those areas where building had already taken place were marked by the very small size of the units of holding, often subdivided. Between these settled areas, plots were considerably larger.

An increased stratification of building types contributed to the survival of the established plot pattern since the bigger houses then being built could be accommodated in the larger plots between the earliest settled areas. Development in Ironbridge between *c.*1820 and 1880 was dominated by the building of substantial single houses – the detached villas – though there were also several formally planned terraces and groups of smaller houses. The villas fitted into small fields which became their gardens: the larger and more valuable the house, the more land it was possible for it to take up and still be an economic investment for the developer.[62] There was a marked contrast between the older areas of settlement, with their crowded and subdivided plots, and those which were developed from empty land in the nineteenth century (fig. 7.8).

The absence of coherently planned urban development reflected a pattern of landholding that was already heavily fragmented. Even where there were large blocks of landholding, a system of leasing had developed which effectively removed the control of building from the hands of the major landowners. Mid-nineteenth-century Broseley, for example, while it was dominated by a single landowner, had a structure of tenants and subtenants which made it analogous to the pattern of landownership and occupancy of Madeley parish where there were large numbers of freeholds.[63] In Madeley Wood and Coalbrookdale, the land held by Joseph Reynolds was dispersed in small plots, and evidence from title deeds for property in these areas suggests that such holdings were commonly developed through leasing: clauses inserted in the leases show that development was expected to be undertaken by tenants.[64]

The survival of a heavily fragmented plot structure combined with the relatively small numbers of new building projects to limit the possibility of overall planning or layouts. A regular building line therefore developed only in the Wharfage–High Street area of Ironbridge. Even here, visual coherence was limited to those blocks in single ownership, where a row of shops might be conceived as an overall design.[65] Generally, new building was undertaken in a series of independent developments, and the

7.8 Ironbridge Hillside, *c.*1890–1900. The centre of Ironbridge was characterised by the intensive use of plots of land. Beyond, single houses in large gardens made different use of the pattern of plots

designed similarity of many different buildings reflects the organisation of building work and the structure of demand rather than the existence of coherent units of development or control. These similarities in building design, even where there was no overall planning or large-scale development, are paralleled in other areas and suggest the currency of specific models of building practice.[66]

But within this apparent coherence of building design there is a detailed hierarchy of building types, and an elaborate series of variations even on superficially similar models of building. Size and number of rooms, the quality of finish and detail in window heads, doorcases and ornamental brickwork were variables within the same basic model. This proliferation of detail suggests close connections between the structure of demand and the organisation of new building work, but these houses were not generally built by their occupiers: the relative scarcity of owner-occupation shows that they were built as speculative development, albeit on a very small scale. Early housing had generally been built by or for first occupiers: inevitably this changed as the housing market developed, and as buildings outlasted their first owners. By the early nineteenth century, although owner-occupancy was not uncommon, there was also a considerable tenant class.[67] The records of title deeds begin to suggest that there was a distinct property-holding class, of traders, entrepreneurs and professionals.[68]

The problem is to reconcile the variety of building types with a speculative building process: in small-scale projects, variations in the amount of money to be invested could have produced something of the variety of forms found in the Gorge. But these variations in the supply side had to correspond with a pattern of demand if the investors were to obtain sufficient return. Without detailed knowledge of rents, prices and the relationship of household structure to the houses themselves, it is difficult to reconstruct the operation of the housing market, but study of census returns does suggest that there was a single household per house in the larger new dwellings.[69] The pattern of variation may therefore also be a product of investors' knowledge of housing needs – of social structure, incomes and currently available housing stock.

All new building must be seen in the context of the whole building stock, and alongside the survival of older buildings. A high proportion of new housing was for a new market – the industrial and commercial middle classes.[70] Census analysis shows that in 1851 the highest proportion of professional, craft and retail occupations were in that part of Ironbridge between Lincoln Hill, The Wharfage/High Street and Church Road: the highest concentration of labourers, miners and ironworkers was in the area to the east of Madeley Hill. At the same time, the census shows intensive use of earlier buildings, with frequent evidence of subdivision and overcrowding. During the nineteenth century, many earlier houses were subdivided to form smaller dwelling units. As in other industrial urban areas, new building had to be economically viable, and housing for the poorest levels of society was not viable.[71] It is this which helps to explain the continuing use of the earliest buildings, some of which were very small and crowded. Nineteenth-century building in the Gorge was characterised by the solidity and sophistication of its design and construction even in quite modest terraces, but studies from other urban areas provide a reminder that it was often the oldest buildings, desperately overcrowded, which became the first slums.[72] The relative values of old and new buildings played an important part in shaping the social landscape of building: it was because there was already a large number of small cottages that new building was dominated by the more expensive villa houses, and that these remained in single occupation.

Trends in housing design and the pattern of building in new development after *c.*1820 were towards the elaboration of distinctions in quality relating to socio-economic structure. Such architectural features as separate entrance halls and decorative façades were used to mark out differences between social groups (see chapter 8). Although there was no large-scale development in the Gorge, by 1850 there was already a sophisticated housing market providing an elaborate hierarchy of qualities.

Above all, nineteenth-century development was shaped by a legacy of buildings from the past. The nuclei of existing buildings were frequently used in new work, and can still be traced as the survival of a ground plan, the incorporation of stone walls in a later brick building, or building round a timber-framed or stone cottage in a new terraced row.[73] Many earlier and larger buildings were subdivided and other earlier buildings adapted to new uses. The pattern of development in the nineteenth century was therefore heavily influenced by its inheritance of earlier buildings, and it was not until the early twentieth century that the replacement of existing buildings was undertaken on any significant scale.

Planning and redevelopment: responses to the building stock

The pre-existing plot structure and building stock formed a vital constituent of development during the nineteenth century, and an inheritance for the public bodies concerned with the regulation of the environment. The public regulatory processes outlined in chapter 6 impacted on a local building industry and housing market which was highly sophisticated and differentiated. There is no evidence to suggest that the market mechanisms for the evaluation of the existing building stock were instantly changed by the system of regulation – condemned houses were not immediately demolished or replaced and in some cases continued in use even after the introduction of public building programmes.

Ultimately, however, here as elsewhere, these building programmes did change standards of house design and organisation in the building industry itself, pressing for larger-scale building firms and the concentration of capital. Housing types that conformed to national models and standards were introduced; the public housing developments and those speculative building projects which adopted similar space standards represented a concentration of development on an entirely new scale, even though they did not substantially modify the existing pattern of plots and holdings; by the 1950s, the existence of public housing was a major element in the re-evaluation and demolition of older buildings.[74] In a parallel process, smaller-scale private development was also designed to replace older property, either literally through demolition, or through a transfer of investment.[75]

All these processes were conditioned as responses to a legacy of buildings: the renewal programmes initiated by Telford Development Corporation also entailed the demolition and replacement of many smaller, older cottages, as well as several of the larger houses which had previously been subdivided. It is only with the recent interest in landscape conservation that investment has again focused on these older buildings, and previous subdivisions have been reversed and small cottages amalgamated.[76]

Both development and redevelopment in the Gorge have continued to be shaped by an inheritance from the first generations of industrialisation: the plot pattern and its survival reflects a structure of landholding in which there was no overall control over development; the organisation of land for industry, and the operation of a property market constrained the shape of settlement, bringing about concentrations of buildings in particular areas; a sophisticated property market and diversified social structure regulated the provision of new buildings and the continued use of old ones. The new buildings provided by the housing market of the nineteenth century, the speculative developments of the twentieth, and the introduction of public housing entailed successive transformations in the use and value of land and building. But the landscape of small plots and the pattern of investment in building established by early industrialisation proved highly adaptable to subsequent patterns of social and economic change.

8

THE MAKING OF AN INDUSTRIAL VERNACULAR

Right on the edge of Madeley Wood are two single-storey houses (fig.8.1 a–b). They are faced like bungalows of the 1950s, but the mix of waste materials – slag, coal-measure sandstone, broken bricks and tiles – used to build them, the untidy collection of additions tacked onto the back, and their position on the top of an old spoil heap betray a different origin: these were once the poor houses of squatters on waste land, rare survivals in this area. Not far away is a larger cottage, with a careful symmetry in its two-unit plan and one and a half storeys, the façade emphasised with decorative Flemish Bond brickwork: this is an example of eighteenth-century industrial vernacular, typical in the Gorge. These houses stand at two extremes of building – on the one hand, the expedient use of whatever materials came to hand, on the other, carefully selected materials and a recognisable architectural aesthetic.

But the primitive building is of a type now rare in the Gorge, while the formally designed cottage is common. The contrast between them highlights the level of architectural and constructional sophistication attained in the majority of houses built in the Gorge, and is emblematic of its development as a prosperous industrial area. The industrial vernacular house – the result of a coherent process of design, extensively used – is itself an important product of an industrialising society, while the simultaneous existence of contrasting modes of building suggests the complex social and economic texture of an industrial community. But industrial vernacular building is a process subject to change, and the relationships between modes of building have likewise been fluid. These patterns of change are integral to a landscape shaped by a dynamic economic and social system, and the relationships and patterns recorded in the design of local buildings are a rich source of evidence for the breadth and depth of industrialisation in the Gorge.

Early industry and the vernacular threshold

In an area celebrated for the early development of its industry, it should not be surprising to find in the Ironbridge Gorge early housing which is specifically industrial, not only in its known connection with an industrial population, but also in the characteristics of its design, construction and use. The brick cottage so typical of the early eighteenth century is arguably such a vernacular building, but is not the first

8.1(a)–(b) Contrasting building processes: two houses in Madeley Wood (M. Worthington)

generation of house-building in the Gorge – and the significance of this new vernacular can be elucidated by consideration of its precursors.

Industrialisation in the area has a longer history than surviving buildings and settlements can themselves document: buildings have survived in significant numbers only from the end of the seventeenth century, although from documentary records we can locate the beginnings of extensive new settlement devoted to industry from the early seventeenth century.[1]

What happened to the earliest houses of the Gorge? Evidence is scanty, but a series of seventeenth-century pictorial maps of Broseley provide some clues as to its distribution and character there.[2] Comparison of these maps with the landscape that survives suggests radical changes of land use and arrangement in some areas, where early buildings were probably completely destroyed, as areas once built over were cleared before later redevelopment.[3] On other plots, houses were apparently rebuilt, even though land use remained apparently unchanged. Surviving seventeenth-century buildings corresponding to those shown on these maps are rare.[4]

The poor survival rate for buildings of the earliest phases of industrial development is important. The existence of first-generation houses which were replaced as soon as circumstances permitted is a phenomenon common to other areas both of rural and of industrial settlement, and, conceptualised as the 'vernacular threshold', it provides an indicator of the local accumulation of wealth and resources, indicative of economic development.[5] Where houses do not survive, the implication is that either the inhabitants did not have the means to build durably, or an increased prosperity later created the desire and wherewithal for a complete rebuilding.[6] Both may have been true in the Gorge.

Rebuilding certainly did not necessarily imply flimsy original construction or the use of poor materials. Representations of buildings on the early maps suggest some quite elaborate construction such as storeyed houses with complex plans; buildings survive where a substantial early timber frame was partially replaced, or encased in brickwork during the eighteenth century. But other short-lived buildings may have used poorer techniques, and the apparent extent of replacement suggests the inadequacy of much early building; many early-eighteenth-century houses in the Gorge contain fragmentary relics of roughly made stone structures. Rebuilding is also likely to be associated with changes in the size and organisation of the house – alongside the larger houses, the early maps also record tiny single-storeyed buildings; the archaeological investigation of one early stone-built cottage (11 Benthall Lane) showed a tiny single-storeyed cottage later improved with the addition of a low attic storey. The resulting dwelling was still tiny, but nevertheless represented an improvement on its predecessor.[7]

By contrast, surviving buildings from the seventeenth century, together with others documented in illustrations and photographs, attain a constructional sophistication, and often a size, which marks them out from the fragmentary traces of many of the earliest structures.[8] They were for the most part substantially built, of large scantling timber or well-coursed rubble; the common framing techniques used (box frame with square panels, arch bracing and queen strut roofs) identify a school of timber construction, suggesting the involvement of experienced building craftsmen. There is a wide social cross-section of buildings, ranging from the large riverside mansions of Madeley Wood and Jackfield to small cottages, which were nevertheless built adequately enough to survive with a viable use.[9] This cross-section elucidates the documentary record of probate inventories that shows such a range, with a prevalence of single-unit houses occupied for the most part by miners. These dwellings, while they were small, were not necessarily primitive.[10]

There were already two contrasting building traditions during the seventeenth century: on the one hand, a professional building process (related to rural vernacular building traditions) which left many substantially made timber and stone houses as its legacy;[11] and, on the other, a more rudimentary form of construction which has survived either not at all or as fragments encapsulated in later buildings. The two crudely built cottages at Madeley Wood (while almost certainly built much more recently) may give some clue as to the nature of the lost generation of houses. The contrast with other buildings in the area, even those built at least a century earlier, highlights the level of architectural and constructional sophistication in the large majority of Gorge houses.[12]

The difference between the two traditions of building is one of access to resources, not of stages of architectural development: primitive buildings that were made without reference to a coherent architectural vocabulary were essentially the self-built houses of the poorest levels of society. This primitive building, which exists outside the mainstream of architectural development, self-building with minimum investment in materials and craft, is paralleled in other industrialising areas where early building was very primitive. In West Yorkshire, for example, an observer of the early nineteenth century noted that workers' houses 'appear to have been single apartments without chambers, open to their thatched roofs, and supported upon crooks without chimneys, without pavement, low, confined, dark and comfortless', while in 1820 James Loch said of Ketley, Shropshire, that 'this township is occasioned by a numerous mining population whose wretched huts have been erected from time to time on the waste'.[13] Squatter settlements were characterised by building of this general type, and there are numerous references to rural labourers living in similar constructions into the nineteenth century.[14] The fact that this type of building was very rare in the Gorge after the late seventeenth century marks out its prosperity as an industrial region.

This pattern of development can be paralleled in other areas where the concept of 'vernacular threshold' has been employed to indicate the transition from impermanent or primitive building to 'the sudden emergence of buildings intentionally in permanent materials' or 'the new ability of people to build durably for the first time'.[15] Such a shift is essentially an economic phenomenon, marking a significant accumulation of resources: rebuilding might be a response to perceived new needs as much as the replacement of inherently impermanent construction.[16] In the Gorge, the transition to a more capital-intensive building tradition was under way by a very early date – the late seventeenth century.

The creation of industrial housing types

A dramatic series of developments in local building took place during the early part of the eighteenth century. Settlement grew through new building and rebuilding, but there were also qualitative changes. Houses of the first half of the eighteenth century constitute a recognisable and new housing type: there are large numbers of highly similar buildings, nearly all built of brick, apparently professionally built using coherent techniques and designed according to a model, and as such representing considerable investment (fig. 8.2 (a)–(c)).

The great expansion of settlement during the eighteenth century was dominated by the use of brick in small buildings. Large houses were built intermittently as an individual investment (the houses of mine-owners, ironmasters and barge-owners), but the greatest quantity of building was concerned with small cottage- or house-building (the homes of miners, ironworkers, bargemen). These cottages of the early eighteenth century have an identifiable common form: nearly all are one-and-a-half-storeyed, of single- or two-unit plan, with much detail in common and similar configurations of planning. The degree of similarity suggests that they comprise a model of building practice which remained in use for some 50 years.

Brick had already been used in building in the Gorge and its immediate area: the octagonal tower at Willey was built in 1618; Raddle Hall, Broseley, is dated 1663; and Bedlam Hall was described as a Jacobean mansion before its demolition.[17] But these were houses of the gentry and there is no evidence to suggest that brick was in common use before the early eighteenth century: none of the brick cottages in the north end of Broseley Wood appeared on the map of 1686, while several can be quite precisely dated.

There is evidence for much more extensive use of brick from the early eighteenth century: Dale House and Rosehill were both built *c.*1720. Calcutts House and Belmont House used a similar style in the 1750s. Of the smaller houses, a cottage on Hodgebower is dated 1714, and Tea Kettle Row was built during the years 1735 to 1742. A cottage on Chapel Road, Ironbridge, is dated 1735, while a pair of cottages in Broseley Wood have title deeds which date them back to the 1740s.[18] The great majority of these cottages do not have direct dating evidence, but a typology of their development would suggest that they largely fall into the first half of the eighteenth century, though there is one terrace in Coalbrookdale dated 1764, and a very similar type of cottage in Wellington is dated 1773. These brick cottages represent a typological development of the stone or timber houses of the seventeenth century – they employ common elements but in a much more formal and controlled way.

This phase of building, dated to the first half of the eighteenth century, marked a transformation of building no less dramatic than that of the origins of permanent building during the seventeenth century. It represents the beginnings of an industrial building tradition: building in greater numbers, building in groups, the adoption of brick, the growing standardisation of design and plan into a recognisable aesthetic. Study of the forms and styles of building enable the process of design and construction

8.2(a)–(c) Eighteenth-century cottage design in the Gorge. During the first half of the eighteenth century, a coherent prototype for domestic building was established, brick-built and compact in plan. Single- and two-unit dwellings, pairs and terraces were derived from the same model (M. Worthington)

8.2(c)

in traditional buildings to be understood: this in turn throws light on the organisation of society.

Studies of the evolution of building types has suggested that, where tradition rather than academic architecture is dominant, change is brought about by the adaptation of existing models.[19] Communication about building design involved an appeal to existing models of practice rather than the production of written plans, and this helps to explain the conservatism of building over quite long periods. Thomas Hubka identified a 'folk design method' that used past precedent by disassembling existing forms into their component parts and recomposing them in new forms. Henry Glassie has also suggested that this recomposition was bounded by rules more or less strictly adhered to.[20] The development of building in the Gorge is also traditional in this sense, and the gradual adaptation of models enables the rules of these so-called traditional processes of design to be traced in detail.

The origins of the eighteenth-century brick cottage type suggest the free adaptation of an existing rural building tradition: the planning of the brick cottages, and their sectional form, with the gabled dormers to an attic storey, appear to be derived from the timber and stone cottages which immediately preceded them. There are several examples of literal adaptation, where a previous timber-framed or stone building formed the basis of later extensions to form a small row, and in at least one of these the timber-framed walls were removed, after erection of a brick load-bearing wall to replace them. In Madeley Wood, a number of early brick cottages were built with exposed timber trusses and framed dormers. The elements of the brick cottage – the planning, the attic dormer windows, the projecting chimney stacks – can all be found in timber or stone cottages, suggesting that it was these buildings which were the source for the new type.[21]

But there are important differences between the timber, stone and brick cottages which help to clarify the relationship between these different building types: taken as a group, the brick cottages have a high degree of regularity in their formal arrangement, whereas the surviving or documented timber and stone houses have far less consistency of design, making it difficult to classify them according to a single model of practice. While there are similarities of constructional technique and basic planning, there is far more variation in volume, placing of doorways, chimneys, windows, and in the relationship between these building components.[22] The brick cottages conformed far more closely to a single model, showing a shift to a more controlled vocabulary and form representing a refinement of a vernacular architectural language under the pressures imposed by an extensive building programme.

Occasional examples of cottages built of stone to exactly similar specifications and probably contemporary with them – stone houses on The Wharfage, Severnside, and Chapel Lane, Ironbridge – show that, while the adoption of brick was important, it was not just a change of materials, but a new approach to the overall building design which distinguished the eighteenth-century building phase. The same basic type could be built in a number of different materials, a phenomenon also characteristic of building elsewhere, showing that building form was not solely determined by materials and constructional technique.[23]

Designing the industrial house

Most surviving early-eighteenth-century houses are based on a single- or two-unit plan form with a corresponding number of heated rooms with gable-end stacks: the most common arrangement in these has a central (or slightly off-central) doorway in the two-unit form, and a door placed away from the stack in a single-unit plan which may therefore be represented as half of the two-unit type. This almost invariable central or near-central doorway became common practice during the eighteenth-century building phase – earlier cottages often had a doorway at one side of the main elevation.[24] Frequently, the windows of the upper and lower storeys align, creating a kind of symmetry. Again, this was established as common practice in the eighteenth century, but was not applied with any consistency in the earlier phases of building.

In most cases, the units were arranged laterally, though in a few exceptional examples a deep-plan form was adopted, with the second room contained as a wing of the main range (as in Tea Kettle Row, for example). Additional half-rooms were contained in an outshut, or in a rear wing, and occasionally in a parallel rear range, and were sometimes later additions. Primary units were subdivided only rarely. A surviving cottage on Hodgebower, Ironbridge, has a wing added to the front of an existing building, and there may once have been other buildings like this. A true double-pile form was adopted in Dale House and Rosehill, by employing a roof construction with a central well, but this was not taken up in other housing before the nineteenth century.

The record of surviving buildings can be compared with that of probates and wills. In the probate records of four parishes (Benthall, Broseley, Little Wenlock and Madeley) between 1673 and 1759, most houses were listed as having between two and seven rooms, the records showing the minimum number of rooms. The industrial parishes of the Gorge had the highest proportion of one-to-four-roomed houses, implying a predominance of single- or two-unit planning. The greater number of houses (17 from a sample of 48) in the parishes of Madeley and Benthall comprised a principal room, variously called the dwelling-house or the kitchen, and a service room (buttery, brewhouse,); a further 10–11 houses of two-room plan had house and parlour; other houses had three rooms, house, parlour and service (12 houses); nine houses recorded a single room, the house. In all cases, these were chambered over. These records show that there were a number of houses with more than two main rooms, though it is not always clear whether additional rooms were contained within the main range or in a rear extension.[25]

The evidence of surviving buildings suggests that, in most cases, the basic two-unit layout was adhered to, and a multiplication of wings and extensions housed additional rooms. This conformity to a two-unit form was, however, sometimes modified by the addition of an extra unit to accommodate additional family members – but, in almost all cases, subsequent subdivision as two or three independent units reintroduced conformity of plan.[26]

Surviving buildings all have either a single- or a two-unit plan, but there are variations in size and scale. These variations may suggest a hierarchy of means but even very small houses were built with the same basic arrangement of plan as larger ones and according to similar standards of construction: the same detailing and techniques were often used both in small cottages and in larger houses.[27]

Single- and two-unit forms can be considered together as a single model of practice: one is half (or twice) the other. The same model was also used in a small number of larger houses: in Coalbrookdale, there are two houses which date from the first half of the eighteenth century and which are of two storeys with gabled attics in the steeply pitched roofs, differing only in form from the cottages by the existence of an additional storey and a central hall (and so employing elements of both vernacular and polite architectural planning).[28] Much smaller variations in size can be charted in the cottage form, and until recently there were several examples of single-unit but two-storeyed houses with attics at Loadcroft Wharf and Jockey Bank, representing an upward extension of the basic type.[29]

Charting these variations of size and scale enables the typological development of this cottage type to be established. For example, there are some houses which have a precise symmetry, where the upper windows align with the lower, and the doorway is exactly central between them. Even where the doorway is not quite central, symmetry is achieved by the alignment of the windows in almost all the houses of the Gorge. Even some of the smallest houses have carefully symmetrical façades. Rural vernacular traditions tended to be more linear and less compact, and visual symmetry was not always adopted. Instead, the different room values were given external expression. The few buildings in the Gorge where the doorway is not central and the windows do not align represent either an earlier form or one built with a lower investment.[30]

The prevalence of compact and symmetrically planned cottages in the Gorge suggests that this configuration became the prototype: it represents a rationalisation of the earlier building forms, the adoption of an aesthetic which became characteristic of urban and industrial building forms. Comparison with the few stone and timber houses which remain shows that the regularity of this building type was in many ways a novel concept for house design. The sheer scale of new building encouraged the adoption of very similar forms, and the use of brick itself may have helped to foster the adoption of more compact planning, since, unlike timber, it was measured and constructed in very small units. The tendency to compact planning was largely a development of the eighteenth-century brick-building phase. The aesthetic which it represented was consolidated in later development, and the symmetrical façade, associated with the adoption of classical motifs in the nineteenth century, became a symbol of urbanisation.

There was a basic uniformity of design in sectional form as well as in ground planning: all these houses have an attic storey, whether they had one and a half or two and a half storeys, though houses of more than one full storey were always rare. Variations to the form chart a range of means and of techniques – room heights varied and attic dormers were contained wholly within the roof or cut the eaves line.

Single- and two-unit cottages were both built in the early eighteenth century, but pictorial and documentary sources, together with surviving buildings, suggest that the two-unit form was the most common. Its early prevalence is also documented in probate inventories, which record a high proportion of buildings of this form; the engraving of upper Coalbrookdale by Vivares, which provides important evidence for the early housing of the Dale, shows a scatter of detached two-unit cottages with two heated rooms and a central doorway.[31] A high proportion of surviving eighteenth-century cottages have a similar form.

The same rules governing plan and sectional form were applied in the design of both single- and two-unit-plan houses. The larger numbers of two-unit houses surviving may suggest that this was the prototype for this model of building, used as the basis of design for a number of different dwelling forms. The process in which a particular building type could be modified by the subtraction or addition of plan elements was identified by Glassie in his study of folk housing in Middle Virginia, showing how ideas were modified to suit economic and environmental needs, as the same builders who built full houses also built fragmented versions for families with less capital or a smaller piece of land.[32]

In England, too, cottages could exist as complete scaled-down versions of small house plans.[33] In the Gorge, the fact that a single-unit house is often exactly half of a two-unit house, and that both these were used in the development of terraces, gives an indication of a similar process. Similarly, the fact that a basic plan and form could be adopted in houses of quite widely different size is also evidence for the use of a dominant model. This was as true in the nineteenth century as it was in the eighteenth: there are several small classical houses which are exact replicas of each other in all but size. The dominance of a type as model is also apparent in the adaptation of building to smaller plots – in Tea Kettle Row, for example, the rear wings are linked together to form a continuous rear range – and exactly the same process of problem-solving was used in the nineteenth century.

The variations of size and scale of building provide important evidence for the social structure of these eighteenth-century industrial settlements. Most notable is the existence of the small, single-unit cottage. Studies of vernacular architecture in other parts of Britain have noted that the single-unit cottage was proper only to the lowest ranks of society, that it was gradually replaced, and is the last building type to be developed in a permanent form.[34] However, in the Gorge, as in other industrialising regions, this smallest of cottages became more, rather than less, prevalent. The smallest cottages that survive appear to be as well built as their larger neighbours, and many are part of a pair or row, designed with attention to an aesthetic, and as an organised development.

At the same time, the numbers of single-unit cottages were increased as earlier two-unit houses came to be subdivided. This process of subdivision took place throughout the eighteenth century: the majority of two-unit houses comprised pairs of single-unit dwellings by the mid-nineteenth century. Single-unit houses were *par excellence* the homes of industrial workers – when the two-unit plan was adopted again in the early nineteenth century, it signalled the emergence of a new middle class. At the same time, very small cottages continued to be built, and the earlier ones continued to be lived in, as the social structure diversified.[35] The fact that the single-unit form became more rather than less common is a mark of the extension of industrial work and suggests not so much a decline in living standards as the growth of specialist work outside the home, and the emergence of a differentiated social structure.

The large houses built for ironmasters and mine-owners belonged to a rather different architectural tradition: a significant change in architectural model also occurred in the design of the larger house, since traditional planning and construction (timber framing, hall and cross-wing plan types) appropriate to the seventeenth century had been abandoned by the early eighteenth, when Dale House and Rosehill were built. These buildings adopted a very different configuration, involving more massing, building in a compact double-pile form, and with a greater number of storeys, and a highly symmetrical composition. The internal planning reflected the external symmetry, arranged about a central entrance and stair hall, and making a more marked distinction between front and back, public, private and service rooms. This principle of planning was also taken up in some houses of an intermediate size associated with works managers and barge- and mine-owners.[36] This plan and form suggests an urban inspiration, and it is likely that drawings or even a surveyor/architect were used in carrying out the work, which could then have been adapted and copied in some of the lesser houses. This planning tradition was also taken up later in houses of the early nineteenth century, where the presence of a central stair hall was one of the distinctions between the more lavish and expensive buildings and the more modest.

Basic building planning was not seriously modified throughout the eighteenth and the beginning of the nineteenth centuries – all the elements employed in planning the nineteenth-century small house had been available since the early eighteenth century, including the central hall and double-pile plan first adopted in Dale House and Rosehill, the single-unit cottage with an additional half-room, and the continuous use of a two-unit plan.

The grouping of industrial houses

The eighteenth-century cottages were built either as single buildings of one- or two-unit-plan form, or as pairs, or as short terraces, though there were a few less coherently arranged clusters of buildings. For the most part, the pairs and terraces appear simply to take the form of single-unit cottages added together to form a row, but there are examples of more innovative planning. Pairs of cottages were often the result of subdivision of an existing house which made little structural impact, but they were also deliberately designed as an integral block, where the subdivision of space between dwellings involved interlocking rooms or attention to external symmetry – two cottages in Broseley Wood, now in use as one, had a similar interlocking of small half-units in the plan, and a strict external symmetry, until modifications were recently made to the layout. Paired houses were also designed as two quite independent dwellings built to adjoin each other.[37]

A terrace like 8–14 Barratts Hill, Broseley, was merely a string of separate houses (fig. 8.3), but others were conceived as the framework for more complex planning, with interlocking units. Units within a terrace varied in size, even when the terrace was a single development, suggesting that the terraced form was thought of as a collection of quite independent houses. One indication of this, for example, is in the arrangement for the construction of chimneys, which are almost never shared between adjoining houses – where this does occur it almost always denotes a phased construction, the adding-on of a new house. The sharing of chimneys was adopted in the Coalbrookdale Company housing of the later eighteenth century but even after that it was not consistently taken up.

There were some more innovative systems of planning: Tea Kettle Row is one of these. Here, each dwelling comprises two and a half units, but instead of being arranged laterally, as would be usual, they are arranged in a double-pile form, and the individual units of the terrace interlock to create an efficient use of space, while maintaining the external appearance of symmetry. This suggests that this row was formally planned, probably on paper, before construction: it was built by Richard Ford, then the manager of the Coalbrookdale Ironworks, as a small speculation. Exactly this system was also taken up in the later Coalbrookdale Company housing, though using much smaller units.

8.3 Creating a terrace: Barratts Hill, Broseley

There were therefore two distinct systems for terrace design, comprising either the terrace as a string of separate houses or the terrace as a unified whole, which could not be separated into its component parts. Tea Kettle Row and Engine Row show that, even where the terrace was conceived as an integral whole, the individual units could still vary in size (as they do in most of the other terraces) (fig. 8.4). There was no conception of standardisation, in which all units had to be identical.[38]

Discussion of development (see chapter 7) has shown how the arrangement of buildings in groups originated in many cases as a solution to a private need within a family, and was often brought about by the piecemeal expansion of an existing house, rather than by deliberate row-building, although there are also examples of short terraces designed and built as a single development.[39] This development process helps to account for the prevalence of terraces which are merely strings of separate houses. But where this piecemeal building did take place it was most often contained within a formal design, so that terraces did emerge with a strong sense of symmetry and sometimes with internal planning sacrificed to external arrangement, as where a property division cuts a window. This provides further evidence that building design was not *ad hoc*, but involved appeal to commonly held models, implemented by professional builders. As a design process, it could be compared with those processes of adaptation which have been recorded in certain settlements such as Holywell Lane, where additions and changes were not organised according to a coherent set of formal rules.[40] In addition, of course, building in groups or rows immediately implies some sort of speculation or organisation of provision.

8.4 Tea Kettle Row, Coalbrookdale, was planned from the outset as a single building, the planning of individual dwellings subordinated to the design of the whole

The construction of the industrial house

From the beginning of the eighteenth century, building in the Gorge was dominated by the use of brick, which was used for the large mansions and small cottages alike. In rural areas, adoption of brick was usually associated with radical architectural and social changes. Eric Mercer noted that

> most brick houses had better accommodation than their predecessors in the same areas: there might be more rooms, the rooms might be larger, and the house was nearly always higher . . . it seems therefore that where brick houses first appeared, they provided the answer to a new demand. They reflected not a slight improvement in standards but a new style of living for an emergent class of moderately wealthy men.[41]

In the Gorge, use of brick in the eighteenth century also marks a transition in the organisation of society, the accumulation of sufficient resources to invest in building, and it also signals the production of building materials on some considerable scale. It shows the development of professional building practice, clearly evidenced in the use of coherent models, in which the elements of building planning were formalised to a degree notably different from most of the earlier stone and timber houses which survive in the Gorge. These changes can be related to the rapid expansion of settlement, and the necessity for an extensive programme of building work.

The particular use of brick in the Gorge gives evidence for the organisation of design and craft in building. Bricks were regularly bonded and their physical properties were exploited decoratively, both in bonding, where the pattern inherent in Flemish Bond, and the different colours of headers and stretchers, could be used to decorative effect, and in such details as dentilled eaves and decorative sill bands. The use of relieving arches and segmental lintels demonstrates an understanding of the constructional properties of brickwork, confirming that these buildings were the work of professional builders. The use of decorative and constructional detail also provides an indication of the resources invested in building – there are a few cottages which are constructed according to more minimal standards, without dentilled eaves, without segmentally arched lintels.[42]

The existence of a clear body of rules in building planning, construction and decoration demonstrates the role played by professional building craftsmen during the eighteenth century. Here, the evidence of architecture supplements a very meagre documentary record, though the terms of some wills refer to houses 'built or caused to be built', and one includes an elaborate set of instructions about building which shows how knowledge of building was not only a professional responsibility, but was also widely shared:[43] in 1720–1 John Brooke, leaving a cottage to his son, charged him with the duty

> at his own proper cost and charges to well and sufficiently build or cause to be builded for my said wife a good, firm and substantial dwelling house . . . with good and well-burnt bricks . . . to contain 16 foots in length, 13 foots in breadth between the walls, and 12 foots high to the wall from the ground to the wall plate, with good and sufficient fire places, and also a shear to goe the length of the said building and an oven for the use of the same. The roof of the said house and all the timber therein used to be all good sawd timber and to be covered with good and sufficient tile.[44]

The detailed description of the proposed property shows how the owner specified a house type, probably with reference to existing practice, in terms of both its dimensions and its construction, showing something of the techniques of communication; the phrase 'cause to be builded' gives a brief suggestion of a commissioning relationship with a builder.

The early development of industrial housing design in the Gorge, showing the early establishment of a school of building which relied on professional craftsmen, is substantially different from the pattern of development that has been traced in other industrial areas, where the shift from primitive to professional building occurred later, often under the aegis of larger-scale land development.[45] What differentiates the

Gorge is the early date for substantial, good-quality housing for workers, and the fact that this was not provided as a planned, paternalist development. Early housing in the Gorge is superior to that in other areas, and is also better than rural cottages of equivalent size. Single-roomed rural cottages of this date, for example, are very rare survivals. Small size and good construction techniques combined by the mid-eighteenth century to demonstrate an organised process for the design and building of workers' dwellings, showing both the availability of resources for building and the existence of an industry of building-materials and craftsmen.

Innovation and tradition before 1800

It was not until the 1780s that a new type of cottage-building was introduced in work undertaken by the Coalbrookdale Company in Coalbrookdale.[46] The four company terraces all employed the same model of building which, unlike the earlier cottage, had two full storeys, a simpler form with a shallower roof pitch, and a single building range. The company housing took up principles of planning that had first been used in Tea Kettle Row, an interlocking form of units giving an efficient use of space within the terrace block (fig. 8.5 (a)–(b)). But in each of the four blocks the precise configuration of internal space varied in each dwelling unit – the terraces were therefore not based on the repetition of identical units. This was an aspect of terrace design also characteristic of earlier building, but in this case, where the variations were contained within a rigidly uniform exterior, it suggests that the design process began by determining the overall area and form of the row, and that the individual house plans were made to fit within that determined volume.

Other details of the design also show a departure from tradition in the interests of efficient utilisation of space, such as the adoption of corner fireplaces, and the grouping of chimneys between individual units. As in the earlier Tea Kettle Row, the design of this housing was not strictly based on an arrangement of traditional forms.

Other development from about the same time bears only a superficial resemblance to these buildings, using two full storeys, but in other respects it follows the principles of planning which had characterised the earlier housing. One important difference was the scaling down of this planning to build much smaller dwellings.[47] This development should perhaps be associated with those changes in the housing market traced in the previous chapter – the emergence of a large tenant class, and the extension of speculative building.

Two distinct design processes can be identified in the development of housing in the Gorge until the late eighteenth century: one, vernacular or traditional, used and adapted existing forms; the other was probably a literate process, starting with drawings. The latter was at this date associated only with development on a larger scale (the company housing was arranged in terraces of eight or ten units – other contemporary building was rarely in rows of more than three units), with greater resources behind it.

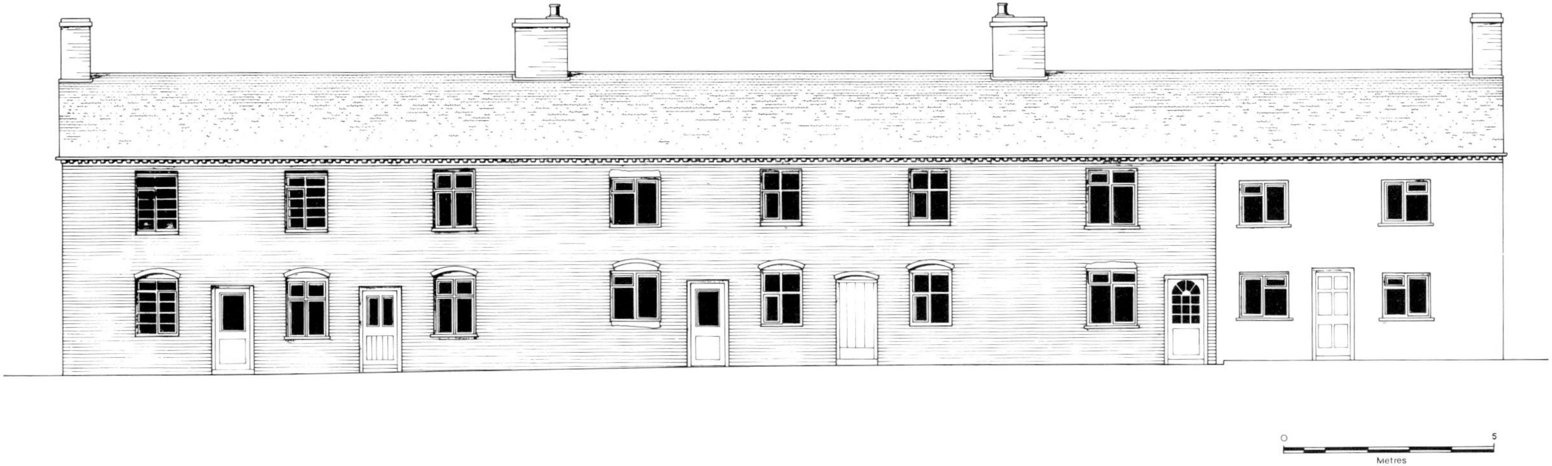

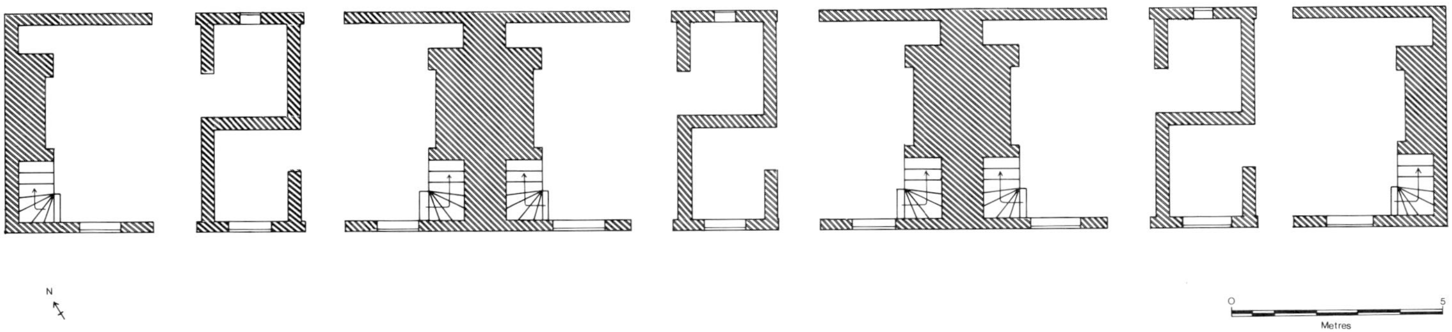

8.5 Charity Row, Coalbrookdale. The Coalbrookdale Company built several terraces of worker's housing during the 1780s. All employed an interlocking plan, first adopted in Tea Kettle Row (Barry Jones). (a) External view; (b) Interlocking plan

Classicism and the nineteenth-century vernacular

In the early nineteenth century, a new phase of house-building was characterised by the adoption of another model house form, the small 'classical' house. This was a two-unit house type, of two full storeys, whose architectural references derived from a classical vocabulary. The central or near-central doorway was contained within an architrave more or less elaborated, sash windows had flat arched heads, sometimes stuccoed in imitation of stone.

The earliest houses of this type were built in Coalbrookdale: a series of maps show the development of Paradise, where houses of this type were being built between *c.*1828 and 1847 (eight in all).[48] A house of the same type on Church Road in Coalbrookdale is probably slightly earlier, and may represent the prototype for this new housing. By 1847, five other houses following the same model had been built along Wellington Road. The model was also used extensively in the development of Ironbridge after *c.*1840 (fig. 8.6 a–b). The earliest dated building using this vocabulary in Ironbridge is a terrace of small dwellings, which, while only one room wide, are evidently derived from the basic classical model: they had been built by 1846.[49] Many of the Ironbridge houses of this type were built after 1847. The latest known example of this housing type was a house on Wellington Road, Coalbrookdale, which was built some time between *c.*1860 and *c.*1870, using the classical vocabulary.[50] A variant of this type was also developed, probably in the 1850s, in which 'picturesque Tudor' detail was substituted for classical – hood moulds and casement windows replaced sashes and flat-arched heads. Building planning remained identical.[51]

Like the cottage form which preceded it, the small classical house also almost always had a two-unit-plan form, usually two rooms of unequal size, with one or two others contained in a rear wing or outshut. Most of the houses were built detached, though sometimes with only a notional separation between them, but there are two terraces in Ironbridge which adopted a single-unit form, and a pair of unequally sized houses built as a single development in Coalbrookdale. The elevation was highly symmetrical, though the central position of the doorway was sometimes displaced in small houses without a central hall. It was invariably flanked by a pair of sash windows, exactly aligned with those upstairs. Only rarely was there an upper window over the doorway.[52] Symmetry and planning were tighter and more compact than in the earlier houses, and room heights were greater with the use of a full second storey.

Mapping the variations in this house type suggests that it was designed according to a vernacular practice similar to the one identified in the building of the eighteenth century, using a prototype adapted to the particular requirements of each development. This is shown especially clearly in those houses where exactly similar planning and detail were used, but transposed to houses of quite different scale (fig. 8.7 (a)–(d)). Similarly, two apparently identical houses forming a pair in Coalbrookdale turn out on examination to have been built at least 20 years apart – one is the exact replica of the other, though very slightly bigger.[53]

By this time, the organisation of building as an industry can be traced in more detail. Not only was the landscape of the Gorge dominated by building-materials

8.6(a)–(b) A new urban style: mid-nineteenth-century houses in Ironbridge and Coalbrookdale (M. Worthington)

8.7(a)–(d) Mid-nineteenth-century houses in Coalbrookdale. The classical or picturesque vocabulary applied to small houses during the mid-nineteenth century used variations in size and detail to produce an elaborate architectural hierarchy (Barry Jones)

industries, with brick- and tile-works proliferating on both sides of the river, offering highly differentiated products, but significant numbers of people were employed in building trades. In 1851, for example, in the immediate area of Ironbridge alone, there were 28 bricklayers, 4 builders, one of whom employed 20 men, several joiners, painters, plumbers and glaziers. Some of the brick manufacturers also acted as builders. The builders were almost always their own architects and only the larger buildings like the Tontine Hotel and the church involved architects who were not themselves builders.[54]

Although the process of design involved in building these houses was the traditional one of adaptation of an existing model, the nineteenth-century classical house was an important departure from the kinds of building which had dominated development during the eighteenth century. In the first place, patterns of use of the earlier houses were changing, and as larger houses were subdivided, there was a growing predominance of the smaller single-unit dwellings: in this context, the classical house represented a new house type even though it was not an innovation in terms of its formal planning. The houses were almost always built in two-unit form as single dwellings; room heights and floor area were larger than earlier buildings; most were detached whereas the tendency throughout the eighteenth century had been towards larger numbers of pairs and terraces of small houses. Detached cottages in single occupation were rare by the early nineteenth century.

A new use was made of architectural decoration and stylistic detail. The classical

8.7(b)

vocabulary had been used in larger houses in the early eighteenth century, when Dale House and Rosehill were built; it was only now taken up in smaller houses. Early use of this vocabulary had been focused on those larger houses built as individual investments, but in the small classical house of the nineteenth century the style was formalised for development on a larger scale, making use of a widely followed model of practice, and building components which were being produced in large quantities by local industries.[55]

In the small cottage of the eighteenth century, similar detailing had been applied across a range of house sizes. In the later classical model, stylistic detail was used consciously to differentiate: large houses of works managers were built according to the same configuration as quite modest small houses but, although the planning was

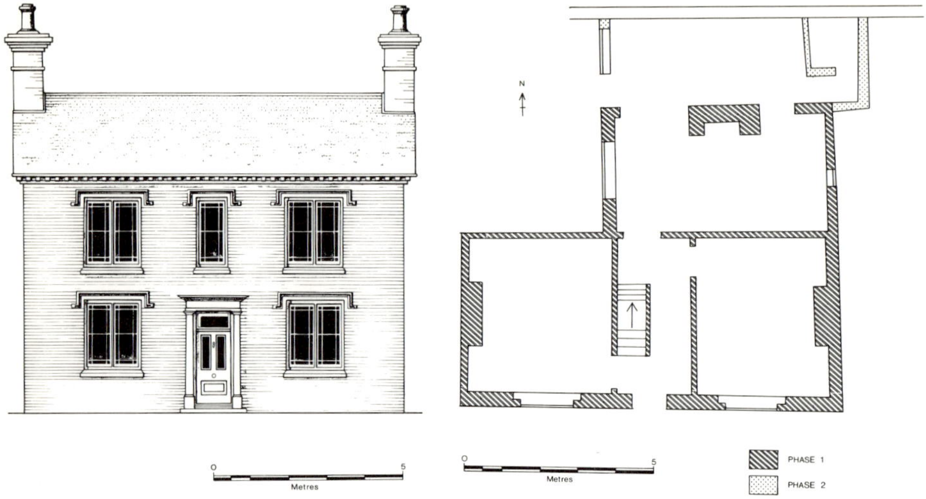

8.7(c)

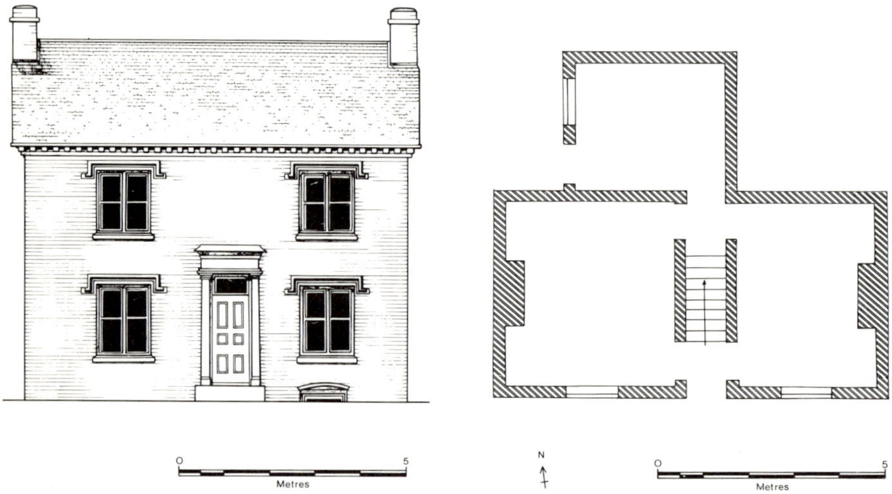

8.7(d)

the same, gradations of social difference were expressed in a fine architectural hierarchy of appropriate decoration. The largest houses had ornate door-cases with fanlights and pediments: window lintels were often decorative cast-iron: the smaller houses more often had minimal timber architraves, and brick-arched heads over windows which imitated stucco with paint.

Above all, the small classical house embodied a new aesthetic. It furthered the use of symmetry, and stressed the façade: the hierarchy of detail reveals the use of casement windows on the back, the façade given special emphasis by the use of decorative brickwork or special facing brick. This aesthetic apparently had a particular connection with urbanism. In Ironbridge, a classical vocabulary was used from the 1780s in the central buildings around the bridge, and its adoption for small houses was part of the expansion of settlement there in the second half of the nineteenth century, consolidating its establishment as a town. Houses of this type were associated with the use of new land for building, and the reshaping of the Gorge as an urban environment. Even in Coalbrookdale, the period of its use coincided with the new provision of communal facilities – school, church, art school and institute.[56]

In Coalbrookdale, where its use originated, the adoption of this house type can be related to a precise social context. The 1830s saw the beginning of a new pattern of industrial organisation within the Coalbrookdale Company, with the closure of the forges and the development of art castings. Associated with this was the emergence of a new class of superior artisan. By 1851, Digby Wyatt noted that the company gave employment to 4000 people, including many with very specialised skills: 'clerks, artists, modellers, carvers, pattern makers, moulders, furnacemen, casters, finishers, smiths, fitters, japanners, painters, gilders, bronzists, and decorators all contribute to the elaboration of that extraordinary series of objects which are constantly issuing from this establishment'.[57] Fogarty noted that 'the skilled workers engaged in the works are a superior class'.[58] The tradition of highly skilled work was not in itself new, but reorganisation and new work practices greatly increased its extent.

At the same time Alfred and Abraham Darby, managing the works jointly from 1830, introduced changes in the regulation and organisation of labour and working conditions: they broke up the old system of semi-autonomous workshops and introduced a new level of foremen and process or department managers.[59] In Ironbridge, too, the new house type can be associated with the development of urban services and trades. A classical vocabulary was adopted exclusively for all commercial development in the wave of building that followed construction of the bridge, and the housing development that followed is closely connected with a class of tradespeople and professionals.

These changes begin to suggest the emergence of a new class structure, parallel to the new stratification of building produced by the development of this larger and more pretentious housing type, alongside the cottages already in existence. While the commercial development of Ironbridge did entail some rebuilding of existing structures according to the new aesthetic, for the most part the small classical houses were an addition to the whole building stock, not a replacement for the earlier cottages: they represented a diversification of the building stock, not merely a change in architectural fashion. This contrasts with the earlier period, where evidence suggests that numbers of buildings were literally replaced. The new model of building was itself highly differentiated in size, setting and degree of architectural elaboration. This differentiation of the building stock had important parallels in the new social hierarchies of changing industrial organisation, commerce and professional services. As part of a differentiated social geography, it was also a major element in the development of the Gorge as an urban area.[60]

The decline of vernacular practice

Ever since the early mine-owners and ironmasters had built mansions for themselves in the seventeenth and early eighteenth centuries, there had been a small but steady volume of building which represented an individual investment, not designed according to a generally held and widely used model, but according to plans and styles specific to each case (fig. 8.8). The building of Dale House and Rosehill (*c.*1720 and 1715–17) marks the first use of academic architecture in the Gorge: a number of other houses followed which employed an individually interpreted classical vocabulary (see, for example, Calcutts House, Belmont House, both 1753, Sunnyside, 1750,

8.8 Distinctive designs: houses in Ironbridge and Coalbrookdale. Individually commissioned houses might aspire to a uniqueness of style.

The White House, probably early nineteenth century). In the nineteenth century, freer styles were adopted in building several large villas in Ironbridge: The Orchard, Southside, Mount Pleasant, The Old Rectory, The Shrubbery were built according to highly individual designs.

But it was not only the largest of the houses which were conceived as individual designs – there were also numbers of other buildings which did not conform to the basic models of practice that were commonly applied. Many of these were built for a specific purpose, so that, for example, while it is possible to identify certain common elements in the design and planning of the many late-eighteenth-century inns that were built in the Gorge, they do not constitute a narrowly interpreted building type.

The distinction between the commonly used vernacular model and the individually designed building is an important one, but it did not mark only a distinction between large houses built as individual investments and smaller cottages built by traditional methods. Not all of the smaller houses in the Gorge were designed according to traditional practice, and not all of the larger houses were built to individual specifications. There is a pair of large late-nineteenth-century villas in Ironbridge that are exact copies of each other, both in planning and in use of materials – both use buff brick and Welsh slate, the latter otherwise rare in the Gorge.[61] A number of smaller houses have quite distinctive designs, suggesting that they were individually commissioned and designed.

During the nineteenth century, the numbers of houses built to individual designs increased as a proportion of the whole housing stock, while the overall volume of building work diminished. The proportion of larger houses that were probably directly commissioned by their owners increased and was especially associated with the growth of Ironbridge, where variety in architecture was seen as an important contributor to its picturesque character.[62] The greater use of individual designs was enhanced by a programme of public buildings, where the process of design and building is occasionally documented: the Ironbridge Police Station, for example, was built by the Borough of Wenlock, following criticism from a government inspector. Plans and a tender for building were obtained from John and Samuel Nevett, Ironbridge builders who were also responsible for the design and construction of the National School.[63] In addition, there were a number of terraced groups as well as individual small houses which did not constitute a recognisable housing type, suggesting that local builders had access to different sources of design, possibly through the use of pattern books.[64] The elaboration of detail in some of these houses may also indicate a level of investment that distinguished them from earlier buildings which were otherwise of a similar size.

The reliance on local vernacular models interpreted by the builders themselves was undermined not only by the growing proportion of individually commissioned houses and public buildings, but also by the introduction of nationally available models: there is one terrace in Coalbrookdale, built *c.*1890, which adopts a plan form common in bylaw housing across the country.[65]

The major building programme of the early twentieth century again took up the use of nationally available models. A public housing programme was initiated in 1919 following the Housing and Town Planning Act of 1919, which gave authority to acquire land and build, as well as to alter, enlarge and repair existing dwellings. Schemes were under way in the area by 1924, but it was not until 1929 that land in Coalbrookdale was acquired for housing development (the product of a second phase of public housing, following recommendations that architects, and not the borough engineers, be used in drawing up plans).[66] The design of these houses followed national imperatives and standards: the Tudor Walters Report, for example, specified that domestic functions should be separated as far as possible, with cooking concentrated in a scullery, and with a separate, indoor bathroom.[67]

These standards were locally interpreted, but on a scale hitherto unknown, as housing provision was planned for a larger community than just the Gorge, and planning was based on an overall perception and interpretation of housing need; it also proceeded in tandem with slum clearance programmes, especially after the Housing Act of 1930. In November 1933, for example, the Madeley Sanitary Committee determined that 23 new houses would be needed to rehouse 118 persons from unfit dwellings.[68] The basic model for building was invariable, though there were differences in size, according to the principal classification into 'parlour'- and 'non-parlour'-type houses. Uniformity was upheld as an ideal: in April 1931, agreeing the tender for a further development on Dale Road, Coalbrookdale, the Madeley Sanitary Committee elected to build pairs of non-parlour houses with bay windows 'because the Council had parlour-type houses adjoining, and for the sake of uniformity the council thought it desirable to adopt them (bays) for the non-parlour type houses'.[69] This uniformity was achieved merely through the exact repetition of plan, form and detail and was not associated with the adoption of a recognisable academic architectural vocabulary.[70]

Uniformity of this exactitude had an element of deliberate social classification, but also reflected changes in the organisation of building practice: the use of architects who were not the builders (Wenlock Borough Council appointed Mr George Riley as its Housing Scheme Architect in January 1925);[71] the necessity to comply with legal requirements regulating structure and services; economies of scale as development was organised in larger units. These factors were also to influence private development, and speculative building in Broseley Wood during the 1950s was characterised by the repetition of identical models.[72] The uniformity of the industrial vernacular which dominated building practice during the eighteenth and nineteenth centuries was reached by quite different means, as an existing model of practice was appealed to, adapted and reinterpreted in each case. But the distinction between building as individual development and building according to a model remains as valid now as it was in the eighteenth century, and, as an indicator of the nature of investment, is an important element in the social hierarchy of settlement.

Conclusion: the formation of an industrial vernacular

The design process that dominated the provision of housing in the Gorge from the early eighteenth to the late nineteenth centuries was a vernacular building tradition of a kind more often associated with rural settlement, but is identifiably the product of an industrialising society. It is vernacular in its use of a type or model of practice, which is constantly adapted and varied. Elements of building were reassembled in different combinations according to clearly defined rules. These rules formed the basis of a communication between client and builder, and in their application they also provide the basis for an understanding of the society that used them.

However, the vernacular process of building design in the Gorge differed from a rural vernacular not only in the scale and organisation of building, but also in the evolution of the formal qualities of the buildings themselves, a pattern of change characterised by the adoption and gradual refinement of highly regular interpretations of building. The interrelationships of design, use and the circumstance of building have helped to shape a distinctively industrial landscape, identifiable through its settlement pattern, its patterns of development and its aesthetic.

House design and patterns of development

The industrial landscape was distinguished by the increased scale of development, as houses built to serve purely individual needs gave place to investment on a larger scale, the building of pairs, terraces and groups of buildings. While there existed groups of buildings with no overall plan or coherent form, the majority of buildings that survived were arranged in more formal groups. A design process specific to the pair or terrace was gradually formulated.

This developed from the stringing together of individually designed houses in a coherent way to the creation of types of design specific to the terrace. The earliest example of this deliberate group design is Tea Kettle Row, which was completed by 1742. In this development, as in others of the early eighteenth century, the size and configuration of the individual units varied: the housing of the Coalbrookdale Company, built in the 1780s, introduced uniform volumes, though individual configuration still varied. By the nineteenth century, the principle of identical units was to dominate building practice, though with the adoption of freer styles in the late nineteenth century varied forms within a group were re-established, but under a different design process.

Building in pairs went through a similar pattern of development, from the joining together of separately conceived dwellings to the creation of deliberate designs that mirrored each other – in the nineteenth century, the concern for symmetry and balance even resulted in the design of pairs imitating a single house.[73]

In all these cases, the more innovative forms continued to employ elements of the traditional vocabulary, but there was a slow shift in the process of design, as the basic type or model, the single house, was displaced as a unit of development, to be replaced by new forms designed according to a different paradigm. But the adoption of new forms was not in this case a complete break with tradition or the introduction of outside ideas. Rather it was the slow adaptation of a building process to suit changing circumstances of development. It was only in the late nineteenth century that styles and types of building that were not traditional in their references were introduced.

House design and use

Perhaps the most important connection between house design and use relates to the ability to afford to build durably in the first place: the design and construction of buildings in the Gorge after the late seventeenth century is clear evidence for the involvement of professional builders. Even the smallest cottages were substantial investments, and it is this shift to professional building which first distinguishes the Gorge as a prosperous industrial region, where most classes had access to sufficient resources either to commission building or to pay the rents of substantially built properties.

In traditional building, the size of house, and the degree of its decoration, corresponded very closely with social position and status. The distribution of building types is therefore an important clue to the understanding of social structure. The scale of building development, the distribution of types, the patterns of their location, the use of architectural detail and style and the age of building are all variables which may be expressive of social status.

In the eighteenth century, the distribution of house types suggests that the basic social divisions were relatively simple – there are large numbers of small single- or two-unit houses, a small number of rather larger dwellings, and a very few large houses. These were closely related to particular places of work. Social distinction was emphasised primarily in distinctions of size, though the larger houses also used an architectural vocabulary which was not adopted for the majority of buildings. These used exactly similar forms, though variations in size and in the quantity of decoration again mapped out differences in access to resources. The higher proportion of surviving small houses after the late seventeenth century is indicative of the expansion of the industrial economy, with a growing class of workers mainly employed outside the home. Although this was a general trend, its timing varied according to very local economic conditions: in Jackfield, for example, a number of inns survived as larger undivided buildings into the nineteenth century, when the decline of the river trade, and the growth of factories, generated a need for small dwellings to replace the larger units of a predominantly domestic economy.

By the early nineteenth century, the social structure (as reflected in the housing stock) was more complex and the introduction of the small classic house represented the adoption of a new housing type, since it was larger, more often built in isolation rather than in groups, and used a distinctive architectural vocabulary. Within this housing type, fine gradations of detail imply a finely graded social hierarchy. In these new houses, architecture was used in a new way to express social distinctions – façades were emphasised with different materials and more elaborate detailing, the scale of decoration of the façade varied with the level of investment in building. The growth and development of an industry of building-materials in the Gorge – the proliferation of brick- and tile-works making products of widely different texture and colour; the introduction of decorative cast-iron building components – multiplied the choice available to builder and client, necessitating a more detailed specification of requirements. These choices remained closely tied to social status and the means available for investment.

At the same time, the earlier, smaller cottages, by now often subdivided to form single-unit dwellings, formed the nuclei of more densely built-up and apparently poorer areas, which were the first to be condemned under the inspection procedures that were developed in the second half of the nineteenth century. The mere survival of older building stock, much of it poorly maintained and in bad condition by the end of the nineteenth century, provided a further dimension to the stratification of society within the Gorge, as those who could afford new houses built them, but the survival of the older and smaller property shows that there was still a market for it. This is confirmed by the fact that very small new houses were also still being built in the second half of the nineteenth century: cottages with tiny one-and-a-half-room plans were built in Church Road, Coalbrookdale, between 1847 and 1881.

Housing and the geography of settlement

The grouping of early eighteenth-century buildings into small clusters in which several cottages were built adjacent to a much larger house, and the connection between this pattern and the organisation of a very local economy, has already been discussed. This pattern survived, but with less resonance in the nineteenth century. Many of the early large houses were subdivided, and a new generation was built beyond the immediate environment of industry itself. At the highest social level, the mansions of the Darby family – Sunnyside and the White House – provide an example of this, though its biggest impact was in the development of Ironbridge, where the slopes of the Gorge were promoted as a picturesque environment.

There is a strong connection between the development of housing in this area and the varied architectural values of individually commissioned private housing. The development of particular architectural styles associated with the function and character of certain areas is also evidenced in the almost exclusive use of a classical vocabulary for the buildings of the centre of Ironbridge, even to the extent of replacing existing buildings to harmonise in style. This process was still important into the twentieth century (and before planning laws provided public sanction for a level of control), and one shop on the Wharfage, for example, was rebuilt as the mirror image of its neighbour.

The interrelationship of these three elements (design, settlement and use) was not only expressed in the architecture of housing: similar distinctions mark commercial building during the nineteenth century, when Ironbridge developed as a small urban centre. Here, the commercial buildings that are grouped round the square use a classical vocabulary articulated to form coherent blocks out of individually owned or occupied units. The design of the commercial premises was specific to their function, containing shops, stores, workshops and dwellings. These were usually three-storeyed, to provide domestic space over the shop itself, and sometimes very large, with additional accommodation for servants, apprentices and assistants. Constraints on design, such as the need for open floor space, encouraged the use of such space-saving features as corner fireplaces and, later, cast-iron columns.[74]

Further away from this commercial centre, shop buildings were either adapted from existing houses or built as modifications of a basic house form, fitting the shop itself into the front room. Sometimes a shop was built as an extension to an existing building, a single-storeyed projection from the façade. Few of these survive now, but are recorded in photographs, while title deeds refer to 'newly erected shops now used as one shop in front of the dwelling', for example. There are even a few small shops, which, while purpose-built, were made in a very rudimentary way, suggesting minimal investment.[75]

In the industrial vernacular of the Gorge, as elsewhere, the architecture and design of buildings was the product of a communication between client and builder, in which there was a close relationship between the architecture and a perception of building use: the architecture was the result of choices about size, degree of decoration and constructional sophistication, which depended on the resources available for building and so on the identity and means of the client or intended occupier.

Several distinct processes of *design* have been identified: at one extreme was self-building, which resulted in rudimentary structures with little explicit architectural quality – apart from an early stage of building, there are few examples of this type of building in the immediate area of the Gorge itself. At the other extreme were the large houses or public buildings which were specially commissioned and designed to individual specifications, probably through a 'literate' process of design involving an architect or the use of a pattern book. But until the later nineteenth century the majority of building work was carried out according to a process of design which may be defined as traditional or vernacular. The basic unit of design, the single house, formed a prototype which was gradually modified to suit changing patterns and scales of development. A new paradigm for design was introduced in the late nineteenth and twentieth centuries, when literate, nationally determined and regulated design processes became standard for almost all building work.

Because of the dominance of certain design methods, the distinction between speculative building and the small houses commissioned by their owner-occupiers is marked primarily at the level of scale: the possibility of economies of scale and the development of more rational solutions. Conversely, the variety of detail observed in the nineteenth-century building stock was not necessarily an index of owner-occupancy.

The developers of building in the Gorge faced a series of choices: location of development, its organisation and the method of design. It is because these choices resulted in a particular range of aesthetic and landscape characteristics that buildings can now be used as evidence for the origins, development and structure of an industrialising community.

9

THE INDUSTRIAL LANDSCAPE TODAY: FROM ANALYSIS TO PROTECTION

The landscape of the Gorge is celebrated as the setting for a series of significant developments in the history of industry and technology: this study has aimed to understand more precisely what constitutes that landscape and how it was developed, setting terms of reference for comparison with other areas. There have been two strands to this inquiry: it has looked at how industrial development took place in the landscape – identifying patterns of development and change in individual industries, in the relationships between industries, as well as in the building and settlement associated with them. It has also considered what this landscape evidence can say about the process of industrialisation, as a sequence of social and economic change.

The central element of the methodology used here was the detailed survey of a small area. This made it possible to range across different types of evidence in great detail. Different classes of evidence could be brought together, and disciplinary boundaries crossed. From this work, it has been possible to explore the use of traditional archaeological techniques such as stratigraphy, context and typology, as well as techniques of analysis developed in vernacular architecture studies.

While the survey sought to test the value of landscape evidence in its own right, it was vital to recognise the limitations of field evidence. The landscape is selective in what it represents. Certain classes of evidence, such as industrial waste, were over-represented in the archaeological record, whilst others (early communities known from documentary records, for example) have completely disappeared. At the same time, poorly documented activities such as limestone working were well represented in the field.

Landscape study is necessarily limited to local events. Economic activities and their organisation are visible in the landscape; markets and external decisions are not. Capital is visible as investment in plant and buildings, but its movement and sources are not. Housing may indicate social structure, and give some evidence for the relationship between industry and labour, but it is not direct evidence of a labour market; how people lived, how they made their livings, the social and economic structure of the household, social mobility, and migration are not visible in the landscape.

There are thus questions which field study will never be able to answer, but every source has its limitations. The landscape remains a primary document in its own right, and deserves the same critical, but respectful, treatment as any other document. The techniques of archaeology and building history used in this study go some way towards this: by studying the relationships between the components of the industrial landscape and how they changed, a detailed understanding of the course of industrialisation has been possible.

By focusing on spatial setting, and the patterns of change over time, it has been possible to look again at the complex, and changing, process of industrialisation. The most striking aspect of this process in the Gorge was its diversity: the iron industry was one of a whole range of industries established long before, and continuing well after, the period of the Industrial Revolution. These industries – and in particular the mining of coal – created an industrial framework on which the spectacular expansion of the iron industry of the 1750s could be based. Equally, the expansion of ceramic industries continued well into the nineteenth century, relying on that same established industrial framework.

In order to understand these industries, it is necessary to understand their mineral base. Coal, clay, limestone and ironstone are each found in a variety of seams, and each seam has different industrial properties. Understanding these properties has to be the first step in charting the pattern of exploitation. It is then necessary to look at the nature of the access to each seam, the changing pattern of demand, and the technology and capital available at any one time to make use of a given seam. The survey also emphasised that mineral resources have to be seen in conjunction with one another, and not just as isolated entities.

Overlying the natural landscape is a human one, created and modified through time. This in itself can be seen as a resource for industry – an existing transport network, power system, and industrialised labour force were all prerequisites for expansion in the mid-eighteenth century. At the same time, such a resource could become a limitation as the ability to expand became hampered by outdated equipment or crowded sites. The constant interplay between the inherited landscape and the pattern of industrialisation shows how important it is to see industry in a chronological framework.

Industrial buildings have been equally useful in investigating this process by marking cycles of innovation, and of 'make-do and mend'. Overall, it has been possible to show how not just a natural but an inherited landscape helped, hindered and influenced the pattern of industrialisation, and that, without an understanding of this inheritance, the process of this industrialisation can never be fully understood.

Study of settlement and housing also reveals the progress of industrialisation in the area. The distinctive pattern and structure of settlement established during the early eighteenth century is evidence for a specialised local economy in which settlements were closely associated with particular patterns of industrial organisation. There was also a qualitative shift in the pattern of building in the early eighteenth century – the use of materials and techniques intended to be permanent, in what is termed the 'vernacular threshold'. These new buildings represented investment substantial enough to be upheld over a long period of time, and give valuable evidence of the organisation of the economy, since they indicate that there was money to invest in building even among miners, trowmen and foundry-workers. The use of professional building craftsmen for small cottages, and the absence of a large property-owning class, or of significant involvement by industrialists, provide clear evidence of local prosperity founded on mining and manufacture.

At the same time, the shift towards smaller dwelling units, and a progressive reduction in the amount of land associated with individual dwellings, suggest a growing dependence on work outside the home: these dwellings were cottages – the homes of people who were not engaged in agriculture, and without land other than a small garden plot. The division of land into many small plots (and among many owners) gives some indication of the distribution of resources, concentrated in few hands only towards the end of the nineteenth century.

The continuing development of the industrial economy, shaped by these new developments, is visible in new patterns of organisation in the landscape: new forms of social stratification are clear from the expanding range of housing types through the later eighteenth century, but especially in the nineteenth. The emergence of a market in land, and of speculative building, is a significant aspect of the economic context of industrialisation which may best be explored through housing evidence.

By the mid-nineteenth century, the limitations of the Gorge were making themselves felt: industries responded by turning to the production of high-value, specialist consumer goods, and the industrial economy shaped by these new developments produced new patterns of organisation in the landscape. By the middle of the century, the Gorge had acquired the character of an integrated urban area, albeit on a very small scale. This urban landscape was shaped by increasingly differentiated patterns of development and land use across the Gorge, with Ironbridge itself as its retail and civic centre. The social hierarchies sustained by changes in industrial and commercial organisation were expressed in the stratification of housing types and their distribution. Subsequent change took place without significantly displacing the structures of settlement and land use established by the later nineteenth century.

This study has defined the Gorge as a particular kind of industrial area, not just by identifying the interrelationship of elements for a structural analysis of industrialisation, but also by examining the course of change over time. Detailed investigation of this local landscape reveals industrialisation as a continuous process rather than an event. The study has shown how, at every stage of development, there was an inheritance of circumstances which established both limits and a framework of possibilities. This inheritance was the structure of landholding, and the organisation of resources in the landscape, but it was also the investment in transport, power systems, and buildings. Periods of new investment, opening up new land, establishing factories on greenfield sites, and major new building programmes stand out from the more usual process of adaptation and compromise.

Not only is the landscape a document, it is also a determinant. In the course of this study, we have been able to identify ways in which the landscape itself constrains and shapes the patterns of historical change. But the landscape which determines is itself the altered landscape which survives. So industry operates within the limits of the land available, on a plot which may owe its form to the clearance of medieval woodland; faced with a complex pre-existing water-power system, the Coalbrookdale Company chose to use new technology to recycle water rather than blow furnaces directly; a key factor in the expansion of the roofing-tile industry during the nineteenth century was the access to deep clays, which was possible as old coal-mines were deepened in the search for new sources. Similarly, many buildings from the late seventeenth and early eighteenth centuries have survived only at the expense of a series of changes. Often, these changes comprise literal adaptation, but may also be less radical changes of use: the movement of houses down a social scale is an example of this. The earliest cottages survived in a diversified social structure as new buildings with higher standards of space brought about a revaluation of the building stock.

Protecting the landscape

A study of the landscape such as this can also contribute to the case for conservation. As more is understood about the historical significance both of particular landscape components and of the structure of the landscape as a whole, so it becomes possible to reassess the criteria used in the evaluation of individual sites and structures, and to develop a more rigorous appraisal of resources.

Any landscape component has some historical significance, but not everything can or should be kept. A degree of selectivity as to what is, and is not, worthy of special protection is always needed. Under the existing planning and legislative framework for protection, seven individual sites within the Gorge have been scheduled as Ancient Monuments – the Darby Furnace, the Iron Bridge, Bedlam Furnaces, Blists Hill Furnaces, the Hay Inclined Plane, Coalport Bridge and Willey Furnaces. Over two hundred buildings have been Listed, including several grade I and II* structures. Most of the Gorge is a Conservation Area in which the demolition of buildings is controlled, as are certain classes of development. Within this area, alterations to existing buildings, and the design of new buildings, must meet strict criteria before they are approved. In 1986, the Conservation Area was designated a World Heritage Site by UNESCO, as part of an attempt to establish an 'effective system of collective protection of the cultural and natural heritage of outstanding universal value'.[1]

Nevertheless, during the period of the survey, the pace of development quickened, with applications both for large-scale working – open-cast mining – and piecemeal new housing development. Despite a relatively high level of statutory protection, change continues to take place rapidly, often taking little account of the historical framework which the survey has begun to identify.

The system of protection itself provides only a partial and unbalanced representation of the scope of landscape resources in the area. The survey highlighted the diversity of industries, but four of the seven Ancient Monuments in the area are iron furnaces. One group of brickworks buildings is Listed, but the remains of the encaustic tile industry at Jackfield have only minimal protection. No limekilns or quarries are Scheduled, and very little protection is given to mining remains. The best protection for mineral remains at present results from the designation of Sites of Special Scientific Interest, where the classification rests on environmental criteria. But woodlands tend to have grown up on old industrial areas, which in turn protect the remains from ploughing and development. This link between natural history and industrial archaeology should be more widely acknowledged, and integrated into the management of these sites in order to afford adequate protection to industrial remains in their own right.

The current system of protection is strongly oriented towards individual sites and buildings. Although linear features such as canals and tramway routes can be protected for their whole length, either by scheduling or through Conservation Area status, often these features are treated as isolated fragments which make little sense, while the water-power systems which unite valleys and groups of sites may be culverted, infilled or diverted. The importance of linear and other systems should also receive specific acknowledgement in planning guidelines.

Historic Buildings listing is one of the cornerstones of effective conservation. Early assessments tended to emphasise particular architectural qualities, but attention is increasingly being given to the significance of buildings in the historical development of the local area. Historical criteria for listing in an area such as the Gorge should assess buildings by reference to their location and development type, rather than by the selection of individual examples of certain architectural styles. The record constituted by the Statutory List should not privilege one period over another, but should adequately represent the changing patterns of development, the changing applications of aesthetic vocabularies over time, and perhaps even typical patterns of alteration and change.

But this study was not just concerned to add to the number of protected sites, monuments and buildings. By focusing attention on the spatial organisation of the landscape, it suggests a way in which historical information can be registered even where literal preservation is not possible. Change within this industrial landscape has previously taken place within a framework of strong continuities in spatial organisation. New development could still be accommodated within that pattern, provided its structure is adequately understood.

It was one intention of the original survey to develop techniques of landscape survey which would be applicable to other areas. Clearly there are many other areas of recognised historical importance which could benefit from the same kind of intensive approach that has been employed here. In particular, areas where there are complex spatial and chronological relationships between land uses could use this approach to gain both a greater understanding of historical processes and a more historically informed basis for landscape management. Above all, the greatest benefit and opportunity of landscape study lies in the identification of previously undervalued historical resources.

NOTES

Abbreviations

BCW: *British Clayworker: The Organ of the Brick and Tile Trades*, London
IGMAU: Ironbridge Gorge Museum Archaeology Unit
IGMT: Ironbridge Gorge Museums Trust
PRO: Public Record Office
SBL: Shropshire Borough Library
SRO: Shropshire Record Office
TDC: Telford Development Corporation
VCH: Victoria County History

1 The landscape as a source of historical evidence

1 W.G. Hoskins, *The Making of the English Landscape*, Harmondsworth, Penguin, 1955.

2 P. Barker, *Techniques of Archaeological Investigation*, London, Batsford, 1982.

3 B. Graslund, *The Birth of Prehistoric Chronology: Dating Methods and Dating Systems in Nineteenth-Century Scandinavian Archaeology*, Cambridge, Cambridge University Press, 1981; D. Clarke, *Analytical Archaeology*, London, Methuen, 1978.

4 Plot survey as a technique to link town plan, building form and land use in urban survey was developed by the historical geographer M.R.G. Conzen in, for example, his surveys of Whitby and Alnwick ('The growth and character of Whitby' in G.H.J. Daysh, ed., *A Survey of Whitby and the Surrounding Area*, Eton, 1958; *Alnwick, Northumberland: A Study in Town Plan Analysis*, Institute of British Geographers Special Publication no. 27, London, Institute of British Geographers, 1960). His work has been the subject of renewed interest and development by a group of historical geographers (J.W.R. Whitehand, 'Background to the urban morphogenetic tradition' and 'Conzenian ideas: extension and development', in *The Urban Landscape*, London, Academic Press, 1981) and stands as a model for all involved in landscape analysis.

5 Plot numbers were allocated to parcels of land in each parish by the Ordnance Survey prior to 1945 after the introduction of the national grid, though they were omitted in dense urban areas (J.B. Harley, *Ordnance Survey Maps: A Descriptive Manual*, Southampton, Ordnance Survey, 1975).

6 Several other maps were available. The Tithe surveys were the earliest large-scale comprehensive coverage of the area, but varied in quality and the selection of detail (R.J.P. Kain and H.C. Price, *The Tithe Surveys of England and Wales*, Cambridge, Cambridge University Press, 1985); the first edition Ordnance Survey maps of the 1880s were only partly available, and the modern Ordnance Survey editions lacked many historical features.

7 M. Biddle, 'Metres, areas, and robbing', *World Archaeology*, vol. 1, 1969, pp. 208–19; E.C. Harriss, *Principles of Archaeological Stratigraphy*, London, Academic Press, 1979.

8 In looking at buildings as the product of particular contexts, this work drew on studies in vernacular architecture, which have explored the connections between buildings and use by emphasising the links between hierarchies of building types and social and economic structure. The approach taken here draws particularly on work by Henry Glassie and others, analysing vernacular architecture in the United States, and principally concerned with identifying processes of design by analysis of characteristic forms and elements of building gathered in a comprehensive area study. A study of housing at Le Creusot (C. Devilliers and B. Huet, *Le Creusot: naissance et développement d'une ville industrielle 1782–1914*) was also valuable: this distinguished between vernacular and polite processes of design, showing how each relates to particular patterns of development. It defined a building type not as a simple classification of architectural elements, but as a concept or model which structured building practice in the past. Many of the studies on vernacular architecture in Britain, drawing out the connections between buildings and a social hierarchy, have also proved invaluable (see, for example, R.W. Brunskill, *Illustrated Handbook of Vernacular Architecture*, London, Faber & Faber, 1978).

9 A comprehensive bibliography of historical sources for the Ironbridge Gorge is to be found in this volume (pp. 233–45).

2 The industrialisation of the Gorge

1 Barrie Trinder, *The Most Extraordinary District in the World*, 2nd edn, Chichester, Phillimore, 1977; John Randall, *Broseley and its Surroundings*, Madeley, Salopian & West Midland Office, 1879; John Randall, *A History of Madeley*, Shrewsbury, Salop County Library, 1880, reprinted 1975; see also accounts in the *Shrewsbury Chronicle* of 1859 and VCH 1908; R.A. Mott, 'Abraham Darby I and II and the coal-iron industry', *Transactions of the Newcomen Society*, vol. 31, 1987 pp. 271–287; Arthur Raistrick, *Dynasty of Ironfounders*, London, Longmans, 1953; Stuart Smith, *A View from the Iron Bridge*, Ironbridge, IGMT, 1979; Catherine Clark, *Ironbridge Gorge*, London, Batsford 1993. The most comprehensive accounts are Barrie Trinder, *The Industrial Revolution in Shropshire*, 2nd edn, Chichester, Phillimore, 1981, and VCH XI and X (forthcoming), on which this chapter is broadly based.

2 H.P.R. Finberg, *Early Charters of the West Midlands*, 2nd edn, Leicester, Leicester University Press, 1971, pp. 197–216.

3 W.E. Wiggins, *Ancient Woodland in the Telford Area*, Telford, Telford Development Corporation, 1986.

4 VCH II.

5 Samuel Parsons' map of Broseley, *c.*1620, SRO 1224/1/32.

6 VCH XI, p. 28.

7 D. Pannett, 'Fish weirs on the River Severn', *Shropshire Newsletter*, vol. 41, 1972, pp. 4–7.

8 At the same time a clear demand for coal can be demonstrated – the 'Chronicles of Shrewsbury' note a fuel crisis, and the excitement of finding coal near Shrewsbury.

9 See chapter 4.

10 SRO 1224/163; M.D.G. Wanklyn, 'John Weld of Willey 1585–1665: an enterprising landowner of the early 17th century', *Journal of the Wilkinson Society*, vol. 3, 1969, pp. 88–99; M.D.G. Wanklyn, 'John Weld of Willey: estate management 1631–1660', *Journal of the Wilkinson Society*, vol. 4, 1970–1, pp. 63–71.

11 Lease quoted in Arthur Raistrick, op. cit., p. 286.

12 PRO Star Chamber 8 310/16, quoted in M.J.T. Lewis, *Early Modern Railways*, London, Routledge, 1970.

13 VCH forthcoming; Parsons' map of Broseley, 1637–8 (fig. 7.1); SRO 1224/1/19.

14 VCH I, p. 454.

15 Field and map evidence (OS map, 1902) shows some pits in the late nineteenth century,

but Benthall was never again seen as a major colliery on a national scale.

16 SRO 1224/163.

17 Barrie Trinder and J. Cox, *Yeomen and Colliers in Telford*, London, Phillimore, 1980; Barrie Trinder and J. Cox, *Miners and Mariners: Probate Inventories for Broseley and Ironbridge*, forthcoming.

18 The shaft is still open and visible; Anon., 'Excavations at Lloyds Coppice Engine', unpublished MS, IGMT Library, 1970; Barrie Trinder, *The Industrial Revolution in Shropshire*, 2nd edn, Chichester, Phillimore, 1981.

19 See chapter 4.

20 Barrie Trinder, *The Industrial Revolution in Shropshire*, pp. 65, 68; see also Telford's account of river levels, in J. Plymley, *A General View of the Agriculture of Shropshire*, London, Board of Agriculture, 1803, pp. 317–33.

21 S. Simpson, *The Agreeable Historian or the Compleat English Traveller*, quoted in Barrie Trinder, *The Most Extraordinary District in the World*, 2nd edn, Chichester, Phillimore, 1988, pp. 15–16.

22 M. Eele, 'An account of making Pitch, Tar and Oil out of a blackish stone in Shropshire, communicated by Mr Martin Eele, the Inventor of it', *Royal Society Philosophical Transactions*, vol. 19, 1697, p. 544.

23 Catherine Clark and Judith Alfrey, *Benthall and Broseley Wood: Third Interim Report of the Nuffield Survey*, Ironbridge, Ironbridge Institute, 1987, pp. 44ff.; John Malam, 'White salt-glazed stoneware manufactured at Jackfield', *West Midlands Archaeology*, vol. 24, 1981, pp. 45–50; see also below, chapters 3 and 5.

24 D.A. Higgins, 'The Clay Pipe Industry of Broseley and Benthall', Ph.D. thesis, University of Liverpool, 1987.

25 Raistrick, op. cit., chapter 4.

26 N. Cox, 'Imagination and innovation of an industrial pioneer: the first Abraham Darby', *Industrial Archaeology Review*, vol. 12, 1990, pp. 127–44.

27 Barrie Trinder and J. Cox, *Miners and Mariners*; see also chapter 6.

28 Joan Thirsk, 'Industries in the countryside', in F.J. Fisher, ed., *Essays in the Economic and Social History of Tudor and Stuart England in Honour of R.H. Tawney*, Cambridge, Cambridge University Press, 1961; reprinted in Thirsk, *The Rural Economy of England*, London, Hambledon Press, 1984, pp. 217–34.

29 Barrie Trinder, 'The development of the integrated ironworks in the eighteenth century', *The Institute of Metals Handbook and List of Members*, London, Institute of Metals, 1988, pp. 217–26.

30 Although leasing often meant that such works lacked permanence.

31 See chapter 4.

32 Roger Edmundson, 'Coalport China Works, Shropshire: a comparative study of the premises and background to their study', *Industrial Archaeology Review*, vol. 3, 1979, pp. 122–45.

33 J. Plymley, *op. cit.*.

34 Barrie Trinder, *Coalport New Town*, Ironbridge, IGMT, 1978.

35 E. Benthall, 'Some xviiith century Shropshire potteries', *Transactions of the Shropshire Archaeological Society*, vol. 55, 1957, pp. 159–70; VCH I, p. 434; A.W.J. Houghton, 'Caughley Porcelain Works near Broseley, Salop', *Industrial Archaeology Magazine*, vol. 5, 1968, pp. 184–92; L. Jewitt, *Ceramic Art of Great Britain*, London, Virtue, 1978, p. 190; SRO 2713/41; see also below chapter 2.

36 A. Jones, *The Coarse Earthenwares*, Ironbridge, IGMAU, 1988.

37 F. Celoria and J. Kelly, 'A post-medieval pottery site and a kiln base found off Albion Square, Hanley, Stoke on Trent, Staffordshire, England', *City of Stoke Museum Archaeological Society Report*, vol. 4, 1973.

38 This section is based on Barrie Trinder, *The Industrial Revolution in Shropshire*, pp. 21–9.

39 M.C. Horton, *Newdale: The North Building*, Ironbridge, Ironbridge Institute, 1989; see also chapter 4.

40 Coalbrookdale Company Minute Book, IGMT library.

41 Raistrick, op. cit.

42 M.C. Horton, *Newdale: An Industrial Township of the Mid-Eighteenth Century*, Ironbridge, Ironbridge Institute, 1989.

43 Barrie Trinder, *The Industrial Revolution in Shropshire*, p. 40; archaeological evidence for stamping and potting has been found at Willey (R. Terry, 'An archaeological survey of the Willey Valley', essay, Ironbridge Institute, University of Birmingham, 1989) and on the Tern at Duncote (Catherine Clark and Amanda Winkworth, *Duncote Farm, Atcham, A Report for English Heritage*, Ironbridge, Ironbridge Institute, 1989) and at Horsehay.

44 Particularly Clod Coal, see below, chapter 3.

45 J.W. Hall, 'Joshua Field's Diary of a tour in 1821 through the Midlands', *Transactions of the Newcomen Society*, vol. 6, 1925–6, pp. 30–2; see also painting by Holmes-Smith, discussed in Stuart Smith, op. cit.

46 Catherine Clark, *Horsehay Ironworks, Shropshire, A Report for Wrekin District Council*, Ironbridge, Ironbridge Institute, 1989.

47 John Randall, *A History of Madeley*; N.M. Dawes, 'A History of Brick and Tile Production in the Coalbrookdale Coalfield', unpublished MS., IGMT library, 1979; A.J. Mugridge, *Brick and Tile Manufactories in the Severn Gorge*, Jackfield, Orchard Press, 1987.

48 See below, chapters 3 and 5; fig. 6.12.

49 See below, chapter 4; VCH I, p. 446; L. Jewitt, op. cit., pp. 184, 190; John Randall, *A History of Madeley*.

50 M. Messenger, *Pottery and Tiles of the Severn Valley*, London, Remploy, 1979, pp. 11ff.; E. Benthall, op. cit.; VCH I, p. 434; Clark and Alfrey, op. cit., pp. 66–71.

51 W.H. Williams, 'Dawley New Town Historical Survey: Industries', unpublished MS., IGMT library, 1964, and local industries card index held in IGMT provide details of local post-war industries.

52 See local directories such as those published by Kelly and Pigot.

53 R.S. Tolley, 'Comments on the changing industrial geography of eastern Shropshire', *West Midlands Studies*, vol. 5, 1972, pp. 1–5.

54 Telford Development Corporation documents held in IGMT library.

55 Telford Development Corporation press release, 1975.

56 Draft Policy Plan for the Severn Gorge, Telford Development Corporation, 1964.

57 E. Thomas, 'Origin of the Ironbridge Gorge Museum Trust', unpublished MS., IGMT library, 1982.

3 Mineral resources in the landscape

1 The local geology is well covered in J. Prestwich, 'On the geology of Coalbrookdale', *Transactions of the Geological Society of London*, 1836, pp. 413–95; John Randall, *A History of Madeley*, Shrewsbury, Salop County Library, 1880, reprinted 1975; T.H. Whitehead *et al.*, *The Country between Wolverhampton and Oakengates, Sheet 153*, Memoirs of the Geological Survey, 1928; B.A. Hains and A. Horton, *British Regional Geology: Central England*, London, HMSO, 1969; Ivor Brown, 'Mineral working and land reclamation in the Coalbrookdale Coalfield', Ph.D. thesis, Leicester University, 1975, pp. 9–27; R.J.I. Hamblin, I. J. Brown and J. Elwood, 'Mineral resources of the Coalbrookdale Coalfield', *Mercian Geologist*, vol. 12, 1969; it is also presented on Geological Survey sheet SJ60SE, 1:10560. Field evidence for mining is best covered in Ivor Brown, 'Underground in the Ironbridge Gorge', *Industrial Archaeology Review*, 1979, pp. 158–69, and in reports of the Shropshire Mining and Caving Club.

2 Peter Toghill and K. Chell, 'Shropshire geology – stratigraphic and tectonic history', *Field Studies*, vol. 6, 1984, p. 59.

3 M.C. Crosfield and M.S. Johnson, 'A study of Ballstone and the associated beds', *Proceedings of the Geological Association*, vol. 25, 1914, pp. 207–9; J. Prestwich, op. cit., pp. 413–95.

4 Catherine Clark and Judith Alfrey, *Benthall and Broseley Wood: Third Interim Report of the Nuffield Survey*, Ironbridge, Ironbridge Institute, 1987.

5 SRO, Cooper and Co., box 188 – a lease of 1806 for Tickwood Ironworks.

6 Quoted in Barrie Trinder, *The Industrial Revolution in Shropshire*, Chichester, Phillimore, 1981, p. 15.

7 Benthall Tithe Map, 1844; *East Shropshire Journal*, 22 October 1800.

8 A date has been cut into the construction of one kiln.

9 Ivor Brown, 'The mineral wealth of Coalbrookdale', *Bulletin of the Peak District Mines Historical Society*, vol. 2, 1965, pp. 255–89; D.A. Holmes, 'The working of Wenlock Limestone', Ironbridge Institute dissertation, 1987.

10 Ove Arup, *Limestone Mines in the Wrekin Area, Report for Shropshire County Council*, privately printed, 1986; Wardell Armstrong, *Limestone Mines in the Wrekin Area: Phase 1 Report for Shropshire County Council*, privately printed, 1984.

11 Stuart Smith, *A View from the Iron Bridge*, Ironbridge, IGMT, 1979; Barrie Trinder, *The Most Extraordinary District in the World*, Chichester, Phillimore, 1988, pp. 22, 48. These superb caverns were accessible until very recently, but have now been filled in because of road collapse.

12 Barrie Trinder, *The Industrial Revolution in Shropshire*, p. 59.

13 Mary Macleod, Michael Stratton and Barrie Trinder, *Llanymynech Hill: An Archaeological and Historical Evaluation*, privately printed for L & R Consultants, Ironbridge, Ironbridge Institute, 1987.

14 One of the difficulties in documenting the industry is that lime-burning was often intermittent and done by itinerant specialists who travelled around a circuit of kiln-owners.

15 John Plymley, *A General View of the Agriculture of Shropshire*, London, Board of Agriculture, 1803, p. 344.

16 Lime-burning, like other mining activities, may have been a second employment. The Post Office Directory of 1856 lists John Patten as lime-burner and barge-owner, although coal-miner and farmer is a more common link.

17 P. Toghill and K. Chell, op. cit., pp. 59–101.

18 The sequence of coals was first published in detail by William Reynolds in J. Plymley, op. cit., p. 57.

19 R. Tylecote, *A History of Metallurgy*, London, The Metals Society, 1976.

20 J. Prestwich, op. cit., p. 259.

21 In 1814 the brick- and tile-maker William Davis of Lloyds Head leased a 'stinking coal foot rid . . . and sulphur coals' from Alexander Brodie (SRO Cooper and Co., box 188).

22 Catherine Clark *et al.*, *Caughley Saggar Works: An Archaeological Evaluation*, Ironbridge, Ironbridge Institute, 1988, fig. 3.

23 Quoted in M.J.T. Lewis, *Early Wooden Railways*, London, Routledge, 1970, p. 99.

24 VCH XI, p. 46.

25 Samuel Parsons' map of Broseley, *c.*1621.

26 Catherine Clark, Amanda Winkworth and Mick Worthington, *Ladywood Bridge Site, Ironbridge, Telford: Excavation and Documentary Research. A Report for Shropshire County Council*, Ironbridge, Ironbridge Institute, 1990.

27 Broseley Hall Estate Map, 1728/65, approximate areas of working can be calculated at 53,125 cubic metres, 9375 cubic metres, and 9687 cubic metres (assuming the coal to be roughly 0.5 m thick). The transition to company working also included limestone – the Madeley Wood Company also took over the Lincoln Hill works.

28 Anon., 'Excavations at Lloyds Coppice Engine, Ironbridge', unpublished MS., IGMT library, 1970.

29 D. Higgins in Michael Trueman, *Archaeology in Ironbridge 1981–1985: Work Carried Out by the Ironbridge Gorge Museum Archaeological Unit*, Ironbridge, Ironbridge Institute, 1988, pp. 44–54; J.S. Leese, 'Old English power plants', *Power*, vol. 36, 1912, pp. 108–9.

30 Remains of a haystack boiler survive at the Blists Hill brick- and tile-works across the canal, and may have been moved from one of these engines.

31 Ivor Brown, 'The Tar Tunnel, Coalport', *Shropshire Newsletter*, vol. 32, 1967, p. 9; Ivor Brown and Barrie Trinder, *The Tar Tunnel*, Ironbridge, IGMT, 1979.

32 Fossil tar was also found at Oilhouse Coppice in Coalbrookdale, but there are no visible remains today. Simpson mentions the extraction of 'British oil' at Coalbrookdale, quoted in Barrie Trinder, *The Most Extraordinary District in the World*, p. 16.

33 Broseley Tithe Map, 1840; table in Catherine Clark and Judith Alfrey, *Jackfield and Broseley: Fourth Interim Report of the Nuffield Survey*, Ironbridge, Ironbridge Institute, 1989, p. 35.

34 Ivor Brown, 'Some mining accidents in the Broseley Field', *Journal of the Wilkinson Society*, vol. 8, 1980, pp. 3–5.

35 F.R. Gameson, 'An 8-man pit – an historic beam engine and the craft of Harry Roberts', *Shropshire Magazine*, 20–1 March 1952.

36 IGMT library, LR/IB/110.

37 J. Prestwich, op. cit., pp. 413–95. Blunt called the seams 'spatous', and 'argillaceous' (in C. Hulbert, *History and Description of the County of Shropshire*, Shrewsbury, 1837, p. 437; printed and published by author).

38 The yield was about 35 per cent iron (T. Parton, *Notes on the Shropshire Coalfield*, London, Colliery Guardian Office, 1868). Brown gives the yield as 665 kg per cubic metre (Ivor Brown, 'Mineral working and land reclamation').

39 Calcining is described by Lewis (quoted in Barrie Trinder, *The Most Extraordinary District in the World*, p. 32).

40 A third type of ironstone was found at Madeley, the New Mine ironstone. It was dug at Styches Pit, and reputedly 6 feet (2 m) thick (T.H. Whitehead *et al.*, op. cit.).

41 In 1793–6, ironstone from Ladywood was valued respectively at £212.10.00 per acre for 'Pinnystone' and £100.17.6 for 'Crawstone' (SRO 515/2/187).

42 R.A. Mott, 'The Coalbrookdale Group Horsehay Works part I', *Transactions of the Newcomen Society*, vol. 31, 1957, pp. 68–70.

43 Ibid., pp. 83–4. Svedestierna noted that 'the Madeley Wood pit iron is some of the best in England. . . . the Craw or Cross stone was a peculiar kind of ironstone and a small quantity of it along with other mines gave the iron that peculiarly soft quality.'

44 Ivor Brown, 'Mineral working and land reclamation', p. 114.

45 J. Prestwich, op. cit., pp. 434–5; Children's Employment Commission, *Children's Employment Commission Report*, British Parliamentary Papers 16, 1842, p. 41.

46 These cottages were owned by the Reynolds family until after the mid-nineteenth century, and one of them is still occupied by descendants of the Brown family for whom they seem to have been originally built.

47 Ivor Brown, 'Underground in the Ironbridge Gorge', p. 163.

48 Ibid., p. 164; Ove Arup, op. cit., p. 18.

49 Details of the rebuilding of these is included in the minutes of the Proprietors of the Iron Bridge (IGMT library).

50 Ivor Brown, 'Mineral working and land reclamation', p. 165.

51 Estimate, reproduced in Neil Cossons and Barrie Trinder, *The Iron Bridge*, Bradford-on-Avon, Moonraker Press and IGMT, 1979, p. 21.

52 Townson, quoted in Barrie Trinder, *The Most Extraordinary District in the World*, p. 53.

53 John Randall, *The Clay Industries Including the Fictile and Ceramic Arts on the Banks of the Severn*, Madeley, The Salopian and West Midland Journal Office, 1877.

54 R.J. Firman and P.E. Firman, 'A geological approach to the study of medieval bricks', *The Mercian Geologist*, vol. 2, 1967, pp. 299–317.

55 Willey Old Hall, Broseley, built by 1618; soft red bricks always predate harder kiln-fired bricks whenever both appear in the same building.

56 SRO 1224/3/503 – a lease of 1734 between Samuel Edwards and Richard Beard.

57 Clark and Alfrey, *Jackfield and Broseley.*

58 N.M. Dawes, 'History of brick and tile production on the Coalbrookdale Coalfield, unpublished MS, IGMT library, 1979; *Architects' Compendium and Catalogue*, 1901; *BCW*, November 1908; Clark and Alfrey, *Jackfield and Broseley*, chapter 6.

59 Arthur Young, 13 June 1776, quoted in Barrie Trinder, *The Most Extraordinary District in the World*, p. 34.

60 *BCW*, May 1894, p. 30.

61 *BCW*, January 1903, p. 332; January 1912, p. 233; November 1886; T. Jones, 'Impressions of the Coalbrookdale area', unpublished MS, IGMT library, n.d.; Clark and Alfrey, *Jackfield and Broseley*, pp. 61–3.

62 *Geological Survey and Museum*, 1920, pp. 148–54; Ivor Brown, 'The mineral wealth of Coalbrookdale', p. 340; Clark and Alfrey, *Benthall and Broseley Wood*, chapter 5.

63 Catherine Clark *et al.*, op. cit.

64 D.A. Higgins, 'The clay pipe industry of Broseley and Benthall', Ph.D. thesis, Liverpool University, 1987.

65 It has also been claimed that the same fine white clay was used for making salt-glazed stoneware at Jackfield (John Malam, 'White salt-glazed stoneware manufactured at Jackfield', *West Midlands Archaeology*, vol. 24, 1981, pp. 45–50), and that pipe clay was the basis of late-eighteenth-century white earthenwares such as Queensware (S. Shaw, *History of Staffordshire Potteries and the Rise and Progress of the Manufacture of Pottery and Porcelain*, Hanley, Simeon Shaw, 1829).

66 SRO Cooper and Co., Broseley Box 189, 7 March 1862, letter from Maws to Harries, the landlord of the Jackfield Pottery, '. . . we do not think you will find any tenant of the potteries would require the red clay for which we ask'; he then goes on to offer to buy the site. The company also offered to buy the clay under Saint Mary's Church, which was to be demolished.

67 Clark and Alfrey, *Benthall and Broseley Wood.*

68 W.A. Smith, 'A Swedish view of the West Midlands in 1802–3', *Journal of West Midlands Studies*, vol. 3, 1970, pp. 45–54.

69 M.D.G. Wanklyn, 'John Weld of Willey 1585–1665: an enterprising landowner of the early 17th century', *Journal of the Wilkinson Society*, vol. 3, 1969, p. 96.

70 Minutes of Proprietors of the Iron Bridge, 8 December 1783.

71 Williams's index, IGMT library.

72 SRO 515/5/234, 515/4/100, 515/2/187.

73 SRO 515/8/232.

74 SRO 515/5/62.

75 Ivor Brown, 'Mineral working and land reclamation', p. 481; D.J. Henkel and R.W. Skempton, 'A landslide at Jackfield, Shropshire in a heavily overconsolidated clay', *Proceedings of the International Conference on the Stability of Earth Slopes*, 1954, pp. 131–7.

76 D. Blake-Roberts and G. Blake-Roberts, 'The results of recent excavations in Coalport, Shropshire', *English Ceramic Circle Transactions*, vol. 2, 1981, pp. 71–80.

4 The human landscape

1 VCH XI, p. 49, on which the next paragraph is also based.

2 The site must have been an improvement on contemporary works such as the ironworks on the Tern where long leats had to be dug to gain sufficient height for a good flow of water (Catherine Clark, *et al., Duncote Farm, Atcham: An Archaeological Evaluation in Advance of the*

A5 By-pass, Ironbridge, Ironbridge Institute, 1989). Although there was a corn-mill at Calcutts, the Severn was rarely used for power as the annual rise and fall was too great to allow a wheel to operate.

3 At Willey one of the furnace pools was created from medieval fish ponds, and at Benthall and Horsehay earlier mill pools had been used.

4 Arthur Raistrick, *Dynasty of Ironfounders*, London, Longmans, Green & Co., 1953, p. 107.

5 'An Afternoon View of Coalbrookdale', *c.*1777, by William Williams (*c.*1740–98) shows a large engine part-way down the valley.

6 R.A. Mott, 'The Coalbrookdale Group Horsehay Works part I', *Transactions of the Newcomen Society*, vol. 31, 1957, pp. 271–87; R.A. Mott, 'The Coalbrookdale Group Horsehay Works part II', *Transactions of the Newcomen Society*, vol. 32, 1959, pp. 43–56.

7 Stuart Smith, 'New light on the Bedlam furnaces, Ironbridge, Telford', *Journal of the Historical Metallurgy Society*, vol. 13, 1979, pp. 21–30.

8 Except during the floods of 1795 when the water ran back up the tunnel (Barrie Trinder, personal communication).

9 Charles Wood, 1754, quoted in Barrie Trinder, *The Most Extraordinary District in the World*, Chichester, Phillimore, 1988, p. 21.

10 Barrie Trinder, *The Industrial Revolution in Shropshire*, Chichester, Phillimore, 1981, pp. 273–4.

11 From an archaeological point of view the new blast furnaces built without extensive water-power systems disappear more readily in the modern landscape. Sites such as Barnet's Leasow, Snedshill, Old Park and Hollinswood were much more ephemeral in landscape terms, and are almost impossible to identify today.

12 Barrie Trinder ('The development of the integrated ironworks in the eighteenth century', *The Institute of Metals Handbook and List of Members*, 1988, pp. 217–26) suggests that charcoal supplies were equally important.

13 Catherine Clark and Judith Alfrey, *Benthall and Broseley Wood: Third Interim Report of the Nuffield Survey*, Ironbridge, Ironbridge Institute, 1987, pp. 77–85.

14 R.B. Matkin, 'John Rennie's diary of a journey through northern England 1784', *East Kent Maritime Trust Historical Study*, no. 2, 1986, p. 24.

15 Hitchcock's map of Benthall 1835, SRO 3956; S. Peskin, 'Early recollections of Coalbrookdale and some of its villages and incidental events', unpublished MS, IGMT library, 1948.

16 J.U. Neff, *The Rise of the British Coal Industry*, London, Routledge, 1932; R.F. Savage and L.D. Smith, 'The waggonways and plateways of east Shropshire', thesis, School of Architecture, University of Birmingham, 1965; M.J.T. Lewis, *Early Wooden Railways*, London, Routledge, 1970.

17 Quoted in Lewis, ibid., p. 96.

18 *The Locomotive Magazine*, 1917; Barrie Trinder, *The Most Extraordinary District in the World*, pp.72–3.

19 Quoted in M.J.T. Lewis, op. cit., p. 241.

20 Catherine Clark and Judith Alfrey, *Jackfield and Broseley: Fourth Interim Report of the Nuffield Survey*, Ironbridge, Ironbridge Institute, 1989, p. 29.

21 N.W. Jones, 'A wooden waggon way at Bedlam Furnace, Ironbridge', *Post-medieval Archaeology*, vol. 21, 1987, pp. 259–62.

22 Barrie Trinder, *The Industrial Revolution in Shropshire*, p. 73.

23 Mark Horton and David de Haan, personal communication.

24 Michael Trueman, *Archaeology in Ironbridge 1981–5: Work Carried Out by the Ironbridge Gorge Museum Archaeology Unit*, Ironbridge, IGMAU, 1988, p. 45.

25 S. Simpson, quoted in Barrie Trinder, *The Most Extraordinary District in the World.*

26 Clark and Alfrey, *Benthall and Broseley Wood*, p. 51; Clark and Alfrey, *Jackfield and Broseley*, p. 136.

27 Barrie Trinder, *The Industrial Revolution in Shropshire*, p. 76.

28 R. Terry, 'An archaeological survey of the Willey Valley', essay, Ironbridge Institute, 1989, gives details of survival to the north of the Gorge.

29 George Young's map of 1788, Boulton and Watt Collection, Birmingham Library.

30 From the evidence of Reynolds' sketchbook.

31 Lease map of Coalbrookdale, 1805, SRO 1681/138; J. and C. Greenwood's map of the county of Salop, 1827, SBL.

32 Barrie Trinder, *Coalbrookdale 1801: A Contemporary Description*, Ironbridge, IGMT, 1979.

33 R. Beale, *The Old Wind: A Preliminary Report*, Ironbridge, IGMAU, 1988.

34 Shropshire Canal Company minutes, quoted in Barrie Trinder, *The Industrial Revolution in Shropshire*, p. 82.

35 Michael Trueman, op. cit., p. 38.

36 S. Duckworth, 'Benthall to Coalport: an interpretive assessment of the south bank of the River Severn in the Ironbridge Gorge', dissertation, Ironbridge Institute, 1988.

37 G. Young's map of 1786, SRO 1681/179.

38 Mentioned in Barrie Trinder, *Coalbrookdale 1801:* the warehouse was taken over by the L&NWR as part of their commitment to retaining the canal–river interchanges, and partly demolished by 1902: Catherine Clark and Judith Alfrey, *Coalport and Blists Hill: Second Interim Report of the Nuffield Survey*, Ironbridge, Ironbridge Institute, 1986, pp. 77–9; Barrie Trinder, *Coalport New Town*, Ironbridge, IGMT, 1978.

39 The history of the Gorge's bridges can be found in A. Blackwall, *Historic Bridges of Shropshire*, Shrewsbury, Shropshire Libraries and the Highways and Transport Department, 1985, with some additional material in Catherine Clark, *Archaeological and Historical Issues Relating to the Proposed New Road Bridge at Ladywood, Ironbridge: A Report for English Heritage*, Ironbridge, Ironbridge Institute, 1989.

40 It may be that these ferries in turn relate to the sites of fords, cleared out of the river in 1741 (Barrie Trinder, *The Industrial Revolution in Shropshire*, p. 69). Ownership, too, was a relevant factor in deciding what land would be available for a bridge, so bridges tended to be built where there was a demonstrable benefit to the landowner.

41 Catherine Clark and Judith Alfrey, *Coalbrookdale: First Interim Report of the Nuffield Survey*, Ironbridge, Ironbridge Institute, 1986, p. 56.

42 *The Engineer*, 27 June 1924, p. 725; Mark Harwood-Little, personal communication.

43 Neil Cossons and Barrie Trinder, *The Iron Bridge*, Bradford-on-Avon, Moonraker Press, 1979, pp. 119–25; A. Blackwall, op. cit., pp. 101–12.

44 Barrie Trinder, *Coalbrookdale 1801.*

45 Catherine Clark, *Archaeological and Historical Issues Relating to the Proposed New Road Bridge at Ladywood, Ironbridge, A Report for English Heritage*, Ironbridge, Ironbridge Institute, 1989, chapter 2.

46 J. Plymley, *A General View of the Agriculture of Shropshire*, London, Board of Agriculture, 1803, p. 273.

47 D.A. Higgins, 'The clay pipe industry of Broseley and Benthall', Ph.D. thesis, University of Liverpool, 1987.

48 Mining map of 1676, SRO 3703/10.

49 Barrie Trinder, *The Industrial Revolution in Shropshire*, pp. 86ff.; Neil Cossons and Barrie Trinder, op. cit., p. 39.

50 Minutes of the Proprietors of the Iron Bridge, 15 September 1775. The contractor, from 1776 at least, was Abraham Darby, who continued to maintain the road.

51 'Furnace cinders 3 tons to the yard in length with 30 loads of ashes to cover the cinders' were used on one route (Minutes of Proprietors of the Iron Bridge, 8 December 1783). Waste industrial material including slag, pipe-waste and tiles was often used for road surfacing (see chapter 2 above). Recent excavations at Duncote have revealed an eighteenth-century road also

paved with iron slag (Catherine Clark and Amanda Winkworth, *Duncote Farm, Atcham*, Ironbridge, Ironbridge Institute, 1990.

52 The distinction between roads and railways is not always clear. Wagonways, gateways, wind-ways, way-boards and horseways are some of the terms used locally on leases and on maps (see detailed discussions in M.J.T. Lewis, op. cit., p. 256, and R.F. Savage and L.D. Smith, pp. 5–7.

53 Minutes of the Proprietors of the Iron Bridge, 17 July 1782.

54 John Marshall, *The Severn Valley Railway*, Newton Abbot, David & Charles, 1989.

55 Catherine Clark, op. cit., p. 11.

56 This must be the explanation for dumps of Coalport waste on the south bank of the river towards Bridgnorth.

5 Industrial buildings in the landscape

1 E. Jones, *Industrial Architecture in Britain, 1750–1939*, London, Batsford, 1985; J.M. Richards, *The Functional Tradition in Early Industrial Buildings*, London, Architectural Press, 1958; A.W. Skempton and H.R. Johnston, 'The first iron frames', *Architectural Review*, vol. 131, 1962, p. 176.

2 The term 'factor' originally referred to an agent or a trader, and a factory was a repository for his goods. It was only in the Health and Morals of Apprentices Act of 1802 that the term was formally recorded as referring to manufacturing buildings (previously 'shop' might have been used). Ure's (1835) definition of the factory system referred specifically to a 'combined operation of many orders of work people in tending... a series of productive machines continuously impelled by a central power'. The definition excluded works in which the mechanism did not form a connected series. The early nineteenth-century usage became enshrined in a series of acts referring to factories, and it was only in 1864 that the legislation was widened, first to cover domestic industry, and later (1867, 1878) to a wide range of industrial premises (B.L. Hutchins and A. Hanson, *A History of Factory Legislation*, London, P.S. King & Son, 1903). In *The Development of the Factory*, London, Cornmarket, 1970, Jennifer Tann adopts the wider definition of the factory as being 'means of production concentrated in one place'.

3 S.D. Chapman, *The Cotton Industry in the Industrial Revolution*, London, Macmillan, 1987, and Jennifer Tann, *The Development of the Factory*, for example, both look at the evolution of the factory in these terms but with particular emphasis on textiles.

4 W.H. Pierson, 'Notes on early industrial architecture in England', *Journal of the Society of Architectural Historians*, vol. 8, 1949, pp. 1–31, sets out a clear distinction between the development of the multistorey factory of light industries, and the complexes of heavier industries such as ironworking, characterised by large, open, single-storey structures of a 'low horizontality'.

5 For example, windmills and non-ferrous metal mines.

6 E. Jones, op. cit., p. 95. Most studies have identified a path whereby industrial architecture draws upon constructional techniques and aesthetics pioneered through engineering (N. Pevsner, *Pioneers of Industrial Design: From William Morris to Walter Gropius*, Harmondsworth, Penguin, 1975), while vernacular traditions were routed by the development of Britain's railway network and the widespread shipment of mass-produced building materials from the 1850s (Alec Clifton Taylor, *The Pattern of English Building*, London, Faber, 1972).

7 The term 'industrial vernacular' is commonly applied to domestic buildings 'whose construction is almost entirely in the hands of a full-time specialist' (B.K. Roberts, *Rural Settlement*, Basingstoke, Macmillan Education, 1987).

8 Eric Mercer, *English Vernacular Houses*, London, HMSO, 1979, p. 196, although Freddie Charles, who was involved in the restoration of the structure, suggests that the building is a combined workshop and dwelling, originally a blacksmith's with a central hearth.

9 VCH XI, p. 48; John Randall, *A History of Madeley*, Shrewsbury, Salop County Library, 1880 reprinted 1975, p. 60.

10 H. Thompson and A. Winkworth, *Little Dawley 1983–85*, Ironbridge, IGMAU, 1988.

11 Coal-measure sandstone was frequently used in Ironbridge in retaining walls or foundations built against earth. For example, the late-nineteenth-century Glynwed engineering building has stone foundations.

12 M.C. Horton, *Newdale: An Industrial Township of the Mid-Eighteenth Century*, Ironbridge, Ironbridge Institute, 1987.

13 The Boring Mill is another industrial building in Coalbrookdale, but the date of the present structure is problematic. There was a boring mill there in 1734, which was rebuilt in 1780 (Arthur Raistrick, *Dynasty of Ironfounders*, London, Longmans, Green & Co., 1953, p.152; Barrie Trinder, *Coalbrookdale 1801: A Contemporary Description*, Ironbridge, IGMT, 1979). The present structure is probably the 1780 building, but could be more recent – it is a brick shed, with steep gabled roof, kingpost trusses with iron ties, and two massive openings, one containing a huge wooden door. It has been built on earlier footings.

14 E. Jones, op. cit.

15 Catherine Clark, *Horsehay Ironworks, Shropshire, A Report for Wrekin District Council*, Ironbridge, Ironbridge Institute, 1989.

16 The building seems to be iron-supported; however, the illustration is somewhat idealised – it suggests four bays rather than three, and arches of a slightly different profile.

17 W. Norris, 'Horsehay Ironworks', unpublished MS, IGMT library, 1876.

18 Ibid.

19 There were obvious advantages to including a crane within a building, but the major limitation at the erecting shops was space. At Horsehay the large travelling cranes were in the open by 1902.

20 Catherine Clark, op. cit.

21 W. Buist, 'A survey of the wing shop, one of the buildings belonging to the Coalbrookdale Company', assignment, Ironbridge Institute, 1990–1.

22 See illustration of Dowlais Ironworks, *c.*1840, in E. Jones, op. cit.

23 W.H. Pierson, 'Notes on early industrial architecture in England', *Journal of the Society of Architectural Historians*, vol. 8, 1949, p. 9.

24 The company used an engraving by W.W. Law entitled 'View of Coalbrookdale' (IGMT CBD59.82).

25 The original function of the building is not known, although by 1884 it was used as the company's commercial offices.

26 The unsuitability of multistorey structures for ironworking is noted in the problems over the Bridgewater foundry, designed so as to be converted to a textile mill if need be (E. Jones, op. cit., p. 100).

27 Ibid., p. 112, although there are similarities to the show warehouses of the textile manufacturers of Leeds, Bradford, Manchester and Batley.

28 D.M. Smith, 'Industrial architecture in the Potteries', *North Staffordshire Journal of Field Studies*, vol. 5, 1965, p. 82; S. Beaver, 'The Potteries: a study in the evolution of a cultural landscape', *Institute of British Geographers Transactions and Papers*, vol. 4, 1964, p. 12.

29 Catherine Clark and Judith Alfrey, *Coalport and Blists Hill: Third Interim Report of the Nuffield Survey*, Ironbridge, Ironbridge Institute, 1987, pp. 66–70.

30 In 1805 this works comprised '3 kilns, hovels, smoke houses, warehouse, stock of earthenware etc.' (SRO 4791/1/5).

31 Three at Jackfield, John Miles' pottery at Ladywood, two potteries at Benthall and one at Coalmoor, Little Wenlock.

32 Larissa Rimmell, 'An archaeological and historical study of the eighteenth-century pottery industry at Jackfield, Shropshire', thesis, Ironbridge Institute, 1983.

33 VCH II, 1967.

34 A.W.J. Houghton, 'Caughley Porcelain Works near Broseley, Salop', *Industrial Archaeology Magazine*, vol. 5, 1968, pp. 184–192; John Randall, *The Clay Industries Including the Fictile and Ceramic Arts on the Banks of the Severn*, Madeley, The Salopian and West Midland Journal Office, 1877, pp. 30–3; illustration in *Salopian Monthly Journal*, April 1875.

35 W. Chaffers, *The New Chaffers Marks and Monograms on Pottery and Porcelain*, London, Reeves & Turner, 1912, p. 766.

36 Thomas Bryan's map of Caughley, 1780, SRO 1224.

37 At the same time Caughley Hall was rebuilt, and Inett Farm remodelled.

38 R.S. Edmundson, 'Coalport China Works, Shropshire: a comparative study of the premises and the background to their study', *Industrial Archaeology Review*, vol. 3, 1979, p. 124.

39 R.S. Edmundson, 'Bradley and Co., Coalport Pottery 1796–1800', *Transactions of the Northern Ceramic Society*, vol. 4, 1981, pp. 127–55; D. Blake Roberts, and G. Blake Roberts, 'The results of recent excavations in Coalport, Shropshire', *English Ceramic Circle Transactions*, vol. 2, 1981, pp. 71–80.

40 R.S. Edmundson, op. cit., p. 126.

41 Reynolds deliberately encouraged industrial development, and soon attracted a timberyard, chainworks, ropeworks, brickworks and bag factory, all of which he said were 'united like links in a chain'. He also built at least 30 houses, and had a vision of a chemical works which was never completed. The area never did grow into a centre for settlement, but was a success as a transport point.

42 It is also radically different from the other row of cottages built at the same time, but not converted – Catherine Clark and Judith Alfrey, *Coalport and Blists Hill: Second Interim Report of the Nuffield Survey*, Ironbridge, Ironbridge Institute, 1988, p. 60.

43 W.H. Pyne, *Microcosm: Or a Picturesque Delineation of the Arts, Agricultures and Industries of Great Britain*, Luton, Luton Museum, 1808, reprinted 1974.

44 SRO 1224/3/503.

45 Arthur Raistrick, op. cit., pp. 87–8.

46 Broseley Tithe Map, 1840. In the 1850s, tiles were always dried under cover, bricks less often so (E. Dobson in F. Celoria, ed., *Journal of Ceramic History*, vol. 5, 1971, p. 42).

47 This impression of almost casual organisation is borne out by the evidence of the Children's Employment Commission of 1862; works relied on independent contractors, who would bring their own children to help and be responsible for paying them. Payment was by piecework, and even the fireman responsible for burning the bricks was paid by results. Patent brickmaking presses were available by 1850 but there is little evidence for the use of machine-made bricks locally. Presses were common in the mass-production of bricks for railways but the railways came late to the area, and even then bricks were supplied from elsewhere.

48 Directory evidence (Pigot's, for example), available by the 1850s, shows a standard range of output, which typically included 'plain and ornamental pressed roof-tiles, ridge, hop and valley tiles in shades of red and brindled', as well as bricks and floor-tiles.

49 Clark and Alfrey, *Coalport and Blists Hill*, p. 60.

50 M. Hammond, 'The Blists Hill Brickworks project', unpublished MS, IGMT library, 1979; N.M. Dawes, 'History of brick and tile production on the Coalbrookdale Coalfield', unpublished MS, IGMT library, 1979; Mark Rowland Jones in Clark and Alfrey, *Coalport and Blists Hill*, pp. 28–35. Elsewhere, isolated buildings survive, including a small office and clay-grinding shed at the Dunge works (R. Pickering, 'The Dunge Brick and Tile Works', thesis, Ironbridge Institute, 1988), a shed at the Rock Works, and a single shed now used as a lorry garage. Most of these buildings have been considerably altered and contain no equipment *in situ*.

51 Broseley Tithe Map, 1840.

52 Further examples can be seen in A.J. Mugridge, *Brick and Tile Manufactories in the Severn Gorge*, Jackfield, Orchard Press, 1987. Recently a small brickworks of this type with its kiln and drying shed intact has been identified at Bourton, to the south of Much Wenlock.

53 In comparison to the 4 million tiles or 1.5 million bricks or a combination of both produced at the original site. (N.M. Dawes, op. cit).

54 One drying shed survives at Jackfield, now a lorry garage; built by 1883, it is now brick with a machine-made decorative tiled roof, pierced by more recent roof lights. Regular brick piers support roof trusses, and there are traces of a boiler base and chimney at the south end. Another shed survives partly intact at the Rock metal recycling works. A small clay-preparation shed associated with the Broseley Tileries stands near the Dunge.

55 *BCW*, 1894 and January 1903.

56 A.T. Herbert, 'Jackfield decorative tiles in use', *Industrial Archaeology Review*, vol. 3, 1979, pp. 146–7.

57 L. Jewitt, *Ceramic Art of Great Britain*, London, Virtue, 1878.

58 The work was under the supervision of Peter Stephen, a Coalport china modeller. Mintons of Stoke threatened legal action, and production ceased (VCH forthcoming, and John Randall, *Broseley and Its Surroundings*, Madeley, The Salopian and West Midland Journal Office, 1879, p. 135.

59 L. Jewitt, op. cit., p. 182.

60 A fire of 1856 destroyed a large order.

61 Most of the major patents – for example, mill machinery for the dry process, the steam blunger, and steam press for dry-pressed tiles – were taken out whilst the Maw brothers were at Benthall.

62 In 1862 the company considered buying the Jackfield Pottery, and had 'lately commenced a distinct branch of manufacture for which the Jackfield Potteries might be made available but the necessary alteration and improvement would be such that no tenant would be inclined' (SRO 1681/189). The statement may refer to the production of majolica.

63 SRO 1681/189.

64 By 1870 dust-pressed tiles were being made. Two colour geometric tiles in patterns of four and eight were found from these partnerships during archaeological investigations.

65 G. Kay, 'Minton, Hollins and Co. The tile factory in Shelton Old Road, Stoke on Trent', dissertation, Ironbridge Institute, 1988; T.A. Lockett, *Collecting Victorian Tiles*, London, Antique Collector's Club, 1982; J. Barnard, *Victorian Ceramic Tiles*, London, Studio Vista, 1972.

66 Excavations by IGMAU confirm that the production of tiles followed pottery at the old kiln; in addition, the 1874 engraving shows altered buildings on the south-western corner.

67 G. Kay, op. cit.

68 John Randall, *Broseley and Its Surroundings*, pp. 146–7.

69 Where all of the machinery was designed and executed by Mr George Kirk of Etruria Ironworks, except the patent clay presses of Messrs Needham and Kite (*The Builder*, 1869).

70 The four kilns in the illustration of 1874 are up-draught, and in a position different from the remains of those that can be seen today.

71 A. Pears, 'Maw and Company: 1900–1910', dissertation, Ironbridge Institute, 1989.

72 SRO 1681/189.

73 And still is. Many examples can be seen in Telford, in the Black Country, the Leeds Street area of Liverpool, and in old textile mills.

74 Apportionments to the Coalbrookdale lease maps of 1805, 1827 and 1838 include references such as 'Old machine house, one tenement', and 'hammer engine, now a tenement', quoted in Catherine Clark and Judith Alfrey, *Coalbrookdale: First Interim Report of the Nuffield Survey*, Ironbridge, Ironbridge Institute, 1986, p. 211.

75 This information comes from the many directories for the area including Pigot (1835, 1838) and Kelly's (1856–1941).

76 See, for example, D. Hey, *Rural Metalworkers of the Sheffield Region: A Study of Rural Industry Before the Industrial Revolution*, Leicester, Leicester University Press, 1972; Marie

Rowlands, *Masters and Men in the West Midlands Iron Trades Before the Industrial Revolution*, Manchester, Manchester University Press, 1975; Paul Collins, *Walsall Leather Industry*, Ironbridge, Ironbridge Institute, 1989.

77 N. Smith, 'An investigation into the commercial life, trades, and craft-industries of nineteenth-century Much Wenlock', dissertation, Ironbridge Institute, 1989.

78 Government initiative also encouraged the Coventry Tool and Gauge Company to move into a set of outbuildings in Madeley in 1941, where they made high-precision tools until 1980. Details of twentieth-century industries come from Williams' index, IGMT library.

79 Kenneth Hudson, *Food Clothes and Shelter*, London, John Baker, 1978, chapter 5.

80 Andrew Ure, *The Philosophy of Manufactures*, London, Bohn, 1835.

81 Jennifer Tann, op. cit.

82 W.H. Pierson, op. cit., notes this in early industrial structures.

6 Landscape of housing: the geography of settlement

1 The Torrington Diaries: Barrie Trinder, *The Most Extraordinary District in the World*, London and Chichester, Phillimore, 1977, p. 34.

2 Daniel Defoe, *A Tour through the Whole Island of Great Britain*, first published 1724–6; this edition, Harmondsworth, Penguin, 1979, pp. 491–2.

3 VCH XI pp. 45–6; above, chapter 2.

4 VCH XI, p. 28; VCH X (draft MS, Broseley); VCH X (draft MS, Benthall).

5 Samuel Parsons' map of Broseley, *c.*1620, SRO 1224/1/32.

6 M.D.G. Wanklyn, 'Industrial development in the Ironbridge Gorge before Abraham Darby', *West Midlands Studies*, vol. 15, 1982, pp. 3–7.

7 VCH XI, p. 28; SRO 2280/2/10.

8 Herefordshire Record Office, Diocesan Probate Records.

9 See below, chapter 8; a detailed inventory of early buildings and discussion of the primary sources available for their study are contained in Catherine Clark and Judith Alfrey, interim reports of the Nuffield Survey, *Coalbrookdale*, *Ironbridge*, *Jackfield* and *Benthall*, Ironbridge Institute, 1985–8.

10 For Bedlam Hall, see H.W. Forrest, *The Old Houses of Wenlock*, Shrewsbury, Wilding & Son, 1922, p. 100; for discussion of the evidence for early housing in this area see Clark and Alfrey, *Ironbridge*, forthcoming.

11 VCH X (draft MS).

12 VCH X (draft MS).

13 See above, chapter 2; Catherine Clark and Judith Alfrey, *Jackfield and Broseley: Fourth Interim Report of the Nuffield Survey*, Ironbridge, Ironbridge Institute, 1989, provides a more detailed discussion of the documentary, archaeological and architectural sources for this area.

14 This settlement pattern is shown in Parsons' map of Broseley, SRO 1224/1/32, and the Broseley Estate Book, Shrewsbury Public Library collection (SPL); it can also be traced through surviving buildings and the photographic records of buildings which survived into the twentieth century, in the archives of SPL and Ironbridge Gorge Museum Trust (IGMT).

15 'The Shropshire Hearth Tax return of 1672', *Shropshire Archaeological and Parish Register Society*, 1949; Adam Crompton's house became 'The Dog and Duck' – along with Lloyds Hall it survived as a row of cottages until *c.*1950. There are photographs of both these houses in the archives of IGMT and SPL; Herefordshire Record Office Diocesan Probate Records.

16 Probate inventories, which often give details of occupation and personal wealth, provide valuable evidence on who lived in the houses of this period; for the organisation of early mining activity, see above, chapter 3.

17 Wanklyn, op. cit.

18 Jeremy Lowe, *Welsh Industrial Workers Housing*, Cardiff, National Museum of Wales, 1977.

19 VCH X (draft MS, Broseley); Wanklyn, op. cit.; J.U. Neff, *The Rise of the British Coal Industry*, London, Geo. Routledge & Sons Ltd, 1932, pp. 308–9.

20 Probate inventories are a valuable source of information about the organisation of the domestic economy: stock-keeping was far less common in the Gorge than elsewhere in the East Shropshire Coalfield at this time (Barrie Trinder, personal communication); see also Barrie Trinder and Jeff Cox, *Yeomen and Colliers in Telford*, London, Phillimore, 1980, pp. 70–2.

21 The process of enclosure and the use of the woodlands are discussed in VCH XI, pp. 40–4.

22 The site of the former village adjacent to Benthall Hall and Church is marked by the remains of house platforms near an area of ridge and furrow – the earliest houses to the west of the parish date from the late seventeenth century, though documentary records suggest that the settlement of miners in the parish was taking place by at least 1630: Catherine Clark and Judith Alfrey, *Benthall and Broseley Wood: Third Interim Report of the Nuffield Survey*, Ironbridge, Ironbridge Institute, 1987, pp. 25–30 and 202; VCH X (draft MS, Benthall).

23 VCH XI, pp. 29 and 36; title deeds in SRO 1681 and 1987 enable the pattern of development of Furnace Bank Farm in upper Coalbrookdale to be traced in detail: they show that, although this farm was largely sold off in small parcels after 1705, these plots were taken from coppice land rather than fields: even in 1780, 35 acres of the original 42 acres of agricultural land remained.

24 VCH X (draft MS, Broseley); Wanklyn, op. cit.

25 Parsons' map, SRO 1224/1/32.

26 See above, chapter 2; Clark and Alfrey, *Jackfield and Broseley*.

27 See above, chapter 2 for the roles of Benthall, Clifford, Brooke *et al.* in the early exploitation of resources.

28 J.D. Chambers and G.E. Mingay, *The Agricultural Revolution, 1750–1800*, London, Batsford, 1966, pp. 96–104; E.L. Jones, 'Agricultural origins of industry', *Past and Present*, vol. 40, 1968, pp. 58–71; Joan Thirsk, 'Industries in the countryside', in F.J. Fisher, *Essays in the Economic and Social History of Tudor and Stuart England*, Cambridge, Cambridge University Press, 1961, pp. 70–88; Marie B. Rowlands, *Masters and Men in the West Midlands Metalware Trades before the Industrial Revolution*, Manchester, Manchester University Press, 1975, p. 5; W.G. Hoskins, *The Midland Peasant*, London, Macmillan & Co. Ltd, 1957, pp. 227–8, 247–76.

29 Detailed discussions of the buildings of this phase are contained in the interim reports of the Nuffield Survey; Trinder, *The Most Extraordinary District in the World*, is an anthology of visitor's descriptions of the Gorge; Stuart Smith, *A View from the Iron Bridge*, Ironbridge, Ironbridge Gorge Museum Trust, 1979, brings together the principal artists' views of the Gorge; the principal deed collections which relate to the development of land in the Gorge are contained in SRO 1681 and SRO 1987: there is a detailed discussion of these sources in Catherine Clark and Judith Alfrey, *Coalbrookdale: First Interim Report of the Nuffield Survey*, Ironbridge, Ironbridge Institute, 1986.

30 Barrie Trinder, 'The open village in industrial Britain', in Marie Nisser, ed., *Industrial Heritage: Transactions of the Third International Conference on the Conservation of Industrial Monuments*, Stockholm, 1981, pp. 375–6; B.A. Holderness, 'Open and close parishes in England in the eighteenth and nineteenth centuries', *Agricultural History Review*, vol. 20, 1972, pp. 126–39.

31 VCH XI, p. 29; Catherine Clark and Judith Alfrey, *Ironbridge: Fifth Interim Report of the Nuffield Survey*, Ironbridge, Ironbridge Institute, forthcoming.

32 Trinder, *The Most Extraordinary District in the World*, p. 19.

33 Smith, op. cit., pp. 14–15.

34 Ibid.; Clark and Alfrey, *Coalbrookdale*.

35 Rosehill and Dale House are both pictured by Vivares in 1758; title deeds recording the building of these houses are SRO 1681/31 and SRO 1681/179; Sunnyside is also pictured by

Vivares; its building is recorded in a memorandum in SRO 1681. Photographed prior to its demolition in 1856. IGMT 1984.6457 and 6440.

36 The Tobacco House and Springhill are both mid-eighteenth century: Woodside House is apparently also substantially of this period, though refronted *c.*1830; Vivares also records another large mid-eighteenth-century house near the site of the present British Legion Club.

37 VCH XI, p. 28; Clark and Alfrey, *Ironbridge*.

38 Benthall Old Vicarage was also built *c.*1700: Clark and Alfrey, *Benthall and Broseley Wood*, pp. 225–6.

39 Acquisition of land in upper Coalbrookdale by the Darby family is documented in title deeds in SRO 1987.

40 W.H.B. Court, *The Rise of the Midland Industries 1600– 1838*, London, Oxford University Press, 1938.

41 Wanklyn, op. cit.

42 The suburban status of Broseley Wood is discussed in Clark and Alfrey, *Benthall and Broseley Wood*; the process by which the settlements of Coalbrookdale, Coalport, Madeley Wood and Jackfield acquired a quasi-suburban status in relation to Ironbridge during the first half of the nineteenth century is discussed below.

43 The influence of tenurial patterns in the riverside communities is discussed in Clark and Alfrey, *Jackfield and Broseley*.

44 Broseley Estate Book; Clark and Alfrey, *Jackfield and Broseley*.

45 J.U. Neff, op. cit., pp. 308–9; VCH X (Broseley draft MS).

46 W.H.B. Court, op. cit.; Marie Rowlands, 'Society and industry in the West Midlands at the end of the seventeenth century', *Midland History*, vol. 4, 1977, pp. 48–60.

47 Wanklyn, op. cit.

48 The gradual diminution of plot sizes as a consequence of land development is discussed below, in chapter 7; in contrast to the Gorge, eighteenth-century lead-miners in the Pennines mixed farming and mining, and scattered hamlets had several acres of meadow and pasture even in the mid-nineteenth century (John Rule, *The Labouring Classes in Early Industrial England*, London, Longman, 1987, pp. 82–4; Ian Blanchard, 'Miners and the agricultural community', *Agricultural History Review*, vol. 20, 1970, pp. 96–106).

49 J. Langton, *Geographical Change and Industrial Revolution: Coalmining in South-West Lancashire, 1590–1900*, Cambridge, Cambridge University Press, 1979.

50 Clark and Alfrey, *Benthall and Broseley Wood*, p. 103; below, chapter 8.

51 R.W. Brunskill, *Traditional Buildings of Britain*, London, Gollancz, 1981, p. 25; M.W. Barley, *The English Farmhouse and Cottage*, London, Routledge & Kegan Paul, 1961.

52 Clark and Alfrey, *Benthall and Broseley Wood*, chapter 3.

53 Clark and Alfrey, *Ironbridge*; Trinder 'The open village in industrial Britain'.

54 This map details 'Warehouse and Tenement, Warehouse and Thomas Finches House'; title deeds to property in upper Coalbrookdale also suggest that house and workplace were sometimes in the same building: SRO 1987/7/3–4 describes a house with an adjacent smith's shop in 1773, the shop converted into a dwelling by 1777; the Pan shops at the lower end of Coalbrookdale were so described on Thomas Slaughter's map of 1775 in IGMT library, but have all the appearance of ordinary domestic buildings of the early eighteenth century.

55 Vivares' engraving of upper Coalbrookdale shows both patterns of development; Clark and Alfrey, *Coalbrookdale*.

56 Joan Thirsk, 'Industries in the countryside'; David Hey, *The Rural Metalworkers of the Sheffield Region: A Study of Rural Industry before the Industrial Revolution*, Leicester, Leicester University Press, 1972; Marie Rowlands, *Masters and Men in the Midlands Iron Trades*; E.L. Jones, 'The agricultural origins of industry'; Pauline Frost, 'Yeomen and metalsmiths: livestock in the dual economy in south Staffordshire, 1560–1720', *Agricultural History Review*, vol. 29, 1981, pp. 29–41.

57 Hoskins, op. cit., chapter 10.

58 Richard Dennis and Hugh Clout, *A Social Geography of England and Wales*, London, Pergamon, 1980, pp. 28 and 46; Fernand Braudel, *Civilization and Capitalism 15th–18th Century*, Vol. 2, *The Wheels of Commerce*, London, Collins, 1983, pp. 321–9.

59 Barrie Trinder, *The Industrial Revolution in Shropshire.*

60 Hans Medick, 'The proto-industrial family economy', in Peter Kriedte, Hans Medick and Jurgen Schlumbohn, eds, *Industrialisation before Industrialisation*, Cambridge, Cambridge University Press, 1981; Franklin Mendels, 'Agriculture and peasant industry in eighteenth century Flanders', *Journal of Economic History*, vol. 27, 1972, pp. 241–61; David Levine, *Family Formation in an Age of Nascent Capitalism*, London, Academic Press, 1977, chapters 2–5, shows how a combination of poor agriculture, the absence of a dominant landowner and unregulated freeholds encouraged the development of industry in the Shepshed area of Leicestershire during the late seventeenth century.

61 Trevor Rowley, *The Shropshire Landscape*, London, Hodder & Stoughton, 1972, pp. 151–2.

62 Swannington Festival Group, *Swannington Village Trail*, 1982, identifies a succession of mining sites in the parish of Swannington, Leicestershire; D.W. Baker, *Swannington Festival Group Development Proposals*, 1984, outlines the chronology of mining development in the parish from shallow bell pits for surface mining to the exploitation of deeper and deeper seams. In the investment by the major landholders, and the existence of specialist miners, this area provides an interesting parallel to the Ironbridge Gorge.

63 W.H.B. Court, op. cit.; David Hey, op. cit.; Marie Rowlands, *Masters and Men in the Midlands Iron Trades.*

64 Buildings of this phase are detailed in the interim reports of the Nuffield Survey.

65 Clark and Alfrey, *Benthall and Broseley Wood*, p. 119; SRO 604.

66 Development in Coalbrookdale is detailed in Clark and Alfrey, *Coalbrookdale*; the Tithe maps for Madeley and Broseley reveal the extent of development at mid-century – the Broseley map in particular records the existence of very tight clusters of buildings, apparently forming groups of virtual back-to-back cottages. A similar pattern of development in Shepshed, Leicestershire, is described in Levine, op. cit., pp. 69–70: 'Mean Cottages, low and narrow... fronting onto the streets or around common yards, and often built in odd shapes to squeeze into odd pieces of land'.

67 Above, chapters 2 and 5; Joseph Farington's sketch of the Gorge provides the best available view of the south bank of the river at this time.

68 Clark and Alfrey, *Jackfield and Broseley.*

69 The Tithe maps for Broseley and Madeley record that subdivision of these larger houses had already taken place.

70 Lease maps; Clark and Alfrey, *Coalbrookdale*, pp. 142–56.

71 Arthur Raistrick, *A Dynasty of Ironfounders*, London, Longmans, Green & Co., pp. 256–63; Lawley, 'Art and ornament in iron: design and the Coalbrookdale Company', IGMT, unpublished manuscript, 1979; Trinder, *The Industrial Revolution in Shropshire*; Clark and Alfrey, *Coalbrookdale*, pp. 154–6.

72 SRO 1987/31; Trinder, *The Industrial Revolution in Shropshire*, p. 127.

73 A map of 1788 shows the area as farm land, and field evidence confirms that, with one exception, there were no buildings in the area at this time. George Young, 'Plan of a Navigable Canal proposed to be made from the Marquis of Stafford's Canal in Lilleshall Parish to Communicate with the River Severn... near Coalbrookdale' (Worcester, 1788), Boulton and Watt Collection, Birmingham City Library; Catherine Clark and Judith Alfrey, *Coalport and Blists Hill: Second Interim Report of the Nuffield Survey*, Ironbridge, Ironbridge Institute, 1986, pp. 89, 91.

74 Trinder, *The Industrial Revolution in Shropshire*, p. 132.

75 Madeley and Sutton Maddock Tithe Surveys show the situation at mid-century; later development traced in Clark and Alfrey, *Coalport and Blists Hill*, pp. 99–100.

76 Ibid., pp. 100–2.

77 This contrast is still visible, though several of the cottages have now been demolished: the Madeley and Sutton Maddock Tithe Maps, and the Ordnance Survey Map of 1883 record this pattern of land use.

78 Trade Directories, which provide a detailed account of retailing, trade and industry after *c.*1820, reveal little expansion or change in the local economy. The population of Coalport remained small. In 1863, it was 'a small place with 400 inhabitants' (Kelly's Directory).

79 John Randall, op. cit, p. 335.

80 Neil Cossons and Barrie Trinder, *The Iron Bridge: Symbol of the Industrial Revolution*, Bradford-on-Avon, Moonraker Press, 1979, pp. 41–6.

81 Anon., *A Description of Coalbrookdale in 1801*, Ironbridge, Ironbridge Gorge Museum Trust.

82 S.D. Chapman, 'Working-class housing in Nottingham during the Industrial Revolution', and M.W. Beresford, 'The back-to-back house in Leeds, 1787–1937', in Stanley D. Chapman, ed., *The History of Working Class Housing: A Symposium*, Newton Abbot, David & Charles, 1971; Martin Daunton, *House and Home in the Victorian City*, London, Edward Arnold, 1983, chapter 2.

83 The Madeley Tithe Map shows the extent of development in 1847. In the central area, little was added after that date; title deeds for a number of properties on the Wharfage and Tontine Hill suggest that the present early-nineteenth-century buildings may have replaced earlier houses (TDC Schedule of Deeds); the evidence for this is discussed in more detail in Clark and Alfrey, *Ironbridge*, forthcoming.

84 A map series, the Tithe Survey of 1847, and the first and second edition Ordnance Survey Maps, enable this development to be traced in detail.

85 The public buildings are discussed in W. Grant Muter, *The Buildings of an Industrial Community*, London, Phillimore, 1979, pp. 10–21.

86 Most of these crowded plots were cleared *c.*1950, but there are a number of photographs in the IGMT archive which record the areas east and west of the bridge.

87 M.R.G. Conzen, 'The morphology of towns in Britain during the industrial era', in J.W.R. Whitehand, ed., *The Urban Landscape: Historical Development and Management*, Institute of British Geographers, Special Publication no. 13, 1981; Richard Dennis, *English Industrial Cities of the Nineteenth Century: A Social Geography*, Cambridge, Cambridge University Press, 1984, chapter 5.

88 Anon., *A Description of Coalbrookdale in 1801*; Neil Cossons and Barrie Trinder, *The Iron Bridge*, pp. 41–6.

89 Charles Hulbert, quoted in Barrie Trinder, *The Most Extraordinary District in the World*, p. 101.

90 Clark and Alfrey, *Ironbridge*.

91 Ibid.

92 The Tithe Apportionment specifies pleasure grounds at the Orchard, Southside and Parsonage; villa development in other towns is also discussed in T.R. Slater 'Family, society and the ornamental villa on the fringes of English country towns', *Journal of Historical Geography*, vol. 4, 1978, pp. 129–44; Leonore Davidoff and Catherine Hall, *Family Fortunes: Men and Women of the English Middle Class 1780–1850*, London, Hutchinson, 1987, pp. 364–6.

93 B.D. Clark and M.B. Gleave, eds, *Social Patterns in Cities*, Institute of British Geographers Special Publication no. 5, 1973; Harold Carter and Sandra Wheatley, 'Some aspects of the spatial structure of two Glamorgan towns in the nineteenth century', *Welsh History Review*, vol. 9, 1978–9, pp. 32–56; Harold Carter, 'Towns and urban systems', R. Lawton, 'Population and society', R.A. Dodgson and R.A. Butlin, eds, *A New Historical Geography of England and Wales*, London, Academic Press, 1978; for a discussion of urban qualities and character, see P.J. Corfield, *The Impact of English Towns 1700–1800*, Oxford, Oxford University Press, 1982; and Whitehand, *Urban Morphology*.

94 A typology of shop buildings is detailed in Clark and Alfrey, *Ironbridge*.
95 Ian Mitchell, 'The development of urban retailing 1700–1815', in Peter Clark, ed., *The Transformation of English Provincial Towns, 1600–1800*, London, Hutchinson, 1984; Roger Scola, 'Retailing in the nineteenth century town', in J.H. Johnson and Colin Pooley, eds, *The Structure of Nineteenth-Century Cities*, London, Croom Helm, 1982; Gareth Shaw and M.T. Wild, 'Retail patterns in the Victorian city', *Transactions of the Institute of British Geographers*, vol. 4, 1979, pp. 278–91; the concepts of Central Place Theory, and Social Area Theory, to which this relates, are discussed in Peter Daniel and Michael Hopkinson, *The Geography of Settlement*, Edinburgh, Oliver & Boyd, pp. 84, 124, 193–8; D.T. Herbert, *Urban Geography: A Social Perspective*, Newton Abbot, David & Charles, 1972; and Richard Dennis and Hugh Clout, op. cit. pp. 14–19.
96 Muter, op. cit. p. 11; VCH XI, p. 64.
97 Ibid., pp. 56–8.
98 Minute books of the Madeley Sanitary Committee, Wenlock Borough Records, detail the building of the police station and court room, and the government report that led to it; Anthony Sutcliffe, 'In pursuit of the urban variable', in Derek Fraser and Anthony Sutcliffe, eds, *The Pursuit of Urban History*, London, Edward Arnold, 1983.
99 Trade Directories and the census enable residents of certain named houses to be traced, although in this area, where there is rarely a coherent street pattern, matching up unnamed houses to the census is not reliable; for Severn House and The Orchard see Muter, op. cit., pp. 61 and 67.
100 John Randall, op. cit., p. 349.
101 Trade Directories suggest that house names like these had become very popular by 1885.
102 The connection between occupational and social hierarchies is discussed in E. Pawson, 'The framework of industrial change', in Dodgson and Butlin, eds, op. cit.; Dennis and Clout, op. cit.
103 For a discussion of Social Area Theory see E. Pawson, 'The framework of industrial change', in Dodgson and Butlin, eds, op. cit.; Dennis and Clout, op. cit.
104 H.P. Dunhill, quoted in *Ironbridge Quarterly*, Ironbridge, 1984.
105 Studies of residential segregation in other towns are made in A.M. Warnes, 'Residential patterns in an emerging industrial town', in Clark and Gleave, op. cit.; Colin Pooley, 'Choice and constraint in the nineteenth-century city', and Richard Dennis 'Stability and change in urban communities: a geographical perspective', in Johnson and Pooley eds, op. cit.
106 Marginal notes in title deeds in SRO 1987 refer to a number of houses 'Pulled down and thrown into the Rope Walk Coppice'.
107 J.N. Tarn, *Five Percent Philanthropy*, Cambridge, Cambridge University Press, 1973, chapters 5 and 7; S. Martin Gaskell, *Building Control: National Legislation and the Introduction of Local Government Bye-Laws in Victorian England*, London, British Association for Local History, 1983.
108 Minute Books of the Madeley Sanitary Committee, Wenlock Borough Records.
109 Ibid.
110 Revised editions of Ordnance Survey Maps, 1927, 1961.
111 Minute Books of the Madeley Sanitary Committee, WBR; VCH XI, pp. 30–1; VCH X (draft MS, Benthall and Broseley).
112 Martin Daunton, 'Public place and private space', in Fraser and Sutcliffe, eds, op. cit., has suggested that there was an ideological element to the opening up of space after the dangerous introspection of the earlier courts; the more open layout of these new building schemes was a characteristic of municipal housing involvement anywhere following criteria set out in the Tudor Walters Report of 1918.

7 Shaping the landscape: processes of development

1 Samuel Parsons' map, SRO 1224/1/32; map of Langley's Tenement, SRO 1224/2/33.

2 This paragraph is based on analysis of plots shown on the Tithe surveys of 1840, 1844 and 1847.

3 The Madeley Tithe Survey shows that most of the land between Lincoln Hill and New Bridge Road was organised as a system of small plots: many of those not taken up for building were in use as gardens.

4 Engraving by Francis Vivares reproduced in Stuart Smith, *A View from the Iron Bridge*, Ironbridge, Ironbridge Gorge Museum Trust, 1979, pp. 14–5.

5 Although Parsons' map of Broseley shows that some plots contained more than one house, the relationship between them suggests either an independent building process, or the creation of a new house through addition to the original dwelling.

6 See, for example, John Burnett, *A Social History of Housing 1815–1970*, London, Methuen, 1978, p. 35.

7 Trevor Rowley, *Villages in the Landscape*, London, Dent, 1978, pp. 142–4, compares settlement at Brierley Hill and Clee Hill with stake claims in Derbyshire; a detailed study of a particular development is provided by Ken Jones *et al.*, 'Holywell Lane: a squatter community in the Shropshire Coalfield', *Industrial Archaeology Review*, vol. 6, 1982, pp. 163–85.

8 VCH IX, pp. 28–9. The sale of small freeholds is detailed in title deeds in SRO 1681.

9 A detailed account of the history of the Broseley Estates is given in VCH X (draft MS); chapter 2 above provides details of the consolidation of landholding by Clifford and Weld; Parsons' map gives information on tenancies, and a rent roll of 1744–5 is reprinted in J. Randall, *Broseley and Its Surroundings*, Madeley, 1879, pp. 60–4.

10 M.D.G. Wanklyn, 'Industrial development in the Ironbridge Gorge before Abraham Darby', *West Midlands Studies*, vol. 15, 1982, pp. 3–7; above, chapter 2.

11 VCH X (forthcoming); above, chapter 2.

12 B.A. Holderness, 'Open and close parishes in England', *Agricultural History Review*, vol. 20, 1972, pp. 126–39; Barrie Trinder, 'The open village in industrial Britain', in Marie Nisser, ed, *Industrial Heritage: Transactions of the Third International Conference on the Conservation of Industrial Monuments*, Stockholm, 1981.

13 For sales of land in the manor of Madeley, see VCH XI, pp. 28–9; the sale of land in Coalbrookdale is documented in title deeds in SRO 1987.

14 VCH X (draft MS); above, chapter 2. The extent of unlicensed building is given in presentments for the manor of the Marsh in 1785 for 'encroachments on the Lord's Waste' (SRO 1224/2/260), but these presentments suggest a way of establishing control without direct involvement in building – the fines would function as rents.

15 Although the type of settlement and plot cannot be directly related to tenure, the nature of building was influenced by it.

16 The limitation on land for new building is especially apparent in 'close' parishes: B.A. Holderness, op. cit.

17 M.W. Beresford and J.G. Hurst, eds, *Deserted Medieval Villages*, Woking, Lutterworth Press, 1971.

18 Some costs are suggested in title deeds: 2 acres of land in Coalbrookdale were sold for £30 in 1738, $3\frac{1}{2}$ acres for £60 in 1743; in 1781, two houses with gardens were sold for £150; in 1745–6, a cottage with a recently added extension, also in Coalbrookdale, was mortgaged for £60: SRO 1987.

19 For Rose Cottage, see Eric Mercer, *English Vernacular Houses*, London, HMSO, 1979, p. 196; unpublished notes prepared by F.W.B Charles, IGMT, 1981; and Thomas Slaughter's 'Plan of Coalbrookdale' in 1753, Museum of Iron, Coalbrookdale; for the Malthouse complex, see G. Young, 'Plan of that Part of Coalbrookdale Bounded by the Lands of John Powis Stanley', 1786, SRO 1681/179; 'Map referring to annexed Lease', 1805, SRO 1681, box 138; and Arthur Raistrick, *Dynasty of Ironfounders*, London, Longmans, Green & Co., 1953,

p. 286. This information is summarised in Catherine Clark and Judith Alfrey, *Coalbrookdale: First Interim Report of the Nuffield Survey*, Ironbridge, Ironbridge Institute, 1986, pp. 36, 106.

20 R. Machin, *The Houses of Yetminster*, University of Bristol Department of Extramural Studies, 1978, pp. 126, 145–55; Machin, 'The great re-building: a re-assessment', *Past and Present*, no. 77, 1977, pp. 33–56.

21 Information on tenancies is taken from a series of title deeds in SRO 1681, 1987 and 1224; Parsons' map provides the evidence for tenancies at will in parts of Broseley Wood.

22 See, for example, S.D. Chapman, 'Workers' housing in the cotton factory colonies, 1780–1850', *Textile History*, vol. 7, 1976, pp. 112–39.

23 This paragraph is based on analysis of distribution and phasing patterns of development during the eighteenth century, drawing on the evidence of standing buildings augmented by pictorial records. Clear examples of the establishment of building lines on plot boundaries are provided by the Jockey Bank and Chapel Road areas of Ironbridge.

24 The Tithe Surveys for the parishes of Benthall, Broseley and Madeley show this coherent pattern of land use in the settled areas – outside these, woodland and waste was recorded, for example, at the Lloyds, Madeley. This is the only area where recognisable 'squatter cottages' still exist.

25 Trinder, *The Making of the Industrial Landscape*, Gloucester, Alan Sutton, 1987, p. 48; Trevor Rowley, *The Shropshire Landscape*, London, Hodder & Stoughton, 1972, pp. 151–2.

26 John Rule, *The Labouring Classes in Early Industrial England 1750–1850*, London, Longman, 1986, pp. 82–5.

27 The pattern of plots can be reconstructed from field and map evidence, the relationships of buildings to plot boundaries giving some indication of plot histories. The siting of buildings may mark out plot boundaries, for example. The Tithe Apportionments enable the pattern of ownership to be superimposed.

28 Richard Dennis, *English Industrial Cities of the Nineteenth Century: A Social Geography*, Cambridge, Cambridge University Press, 1984, chapter 5, describes a similar process of plot subdivision in other urban industrial areas.

29 Trinder, *The Industrial Revolution in Shropshire*, London, Phillimore, 1973, pp. 187–8; Trinder, 'The open village in industrial Britain'.

30 David Levine, *Family Formation in An Age of Nascent Capitalism*, London, Academic Press, 1977, chapters 2–5; Wanklyn, op. cit.

31 This and the following two paragraphs are based on wills and probate inventories in Herefordshire Record Office, Diocesan Probate Records.

32 Ibid.; SRO 1224/3.

33 Occupations are given in wills, probates and title deeds.

34 Trinder, *The Industrial Revolution in Shropshire*, pp. 181–5.

35 Title deeds for Furnace Bank Farm are in SRO 1987.

36 This series of deeds also enables the increasing value of property to be traced, in a period of relatively low inflation: in 1781, two dwelling houses with gardens were sold for £150; a single house was sold for £80 in 1782, the same house fetching £105 in 1790, and £115 in 1797; in 1773 a house and smith's shop were sold for £105, and for £115 seven years later when the shop had also been converted into a dwelling. This former shop had been mortgaged for £80 by 1791.

37 The former settlement at Holywell Lane, recorded in Jones *et al.*, op. cit.; certain areas of the North-West Leicestershire Coalfield display a similar pattern of growth – outlying settlements in the parishes of Swannington and Coleorton, for example.

38 See, for example, 15–17 and 37–39 Hodgebower, 31 Wesley Road, Ironbridge, 26–27 Cobwell Road, Broseley.

39 For examples of early speculative terraces, see 40–42 Quarry Road Broseley, and Tea Kettle Row, Coalbrookdale, both *c.*1740, and 45–47 Wellington Road Coalbrookdale, *c.*1760.

40 The building of Tea Kettle Row is fully documented in SRO 1681/179, which shows

that the land was acquired by Richard Ford, manager of the Coalbrookdale Ironworks, in 1738, and that there were five houses under lease by 1753. Ford himself probably lived in Rosehill House.

41 VCH XI, p. 29. The extent of leasing is revealed in title deeds in SRO 1681 and 1987.

42 Gillian Darley, *Villages of Vision*, London, Paladin, 1978, pp. 17–18.

43 Ibid., pp. 122–37; S.D. Chapman, op. cit.

44 Trinder, *The Industrial Revolution in Shropshire*, pp. 190, 198–9.

45 Clark and Alfrey, *Coalbrookdale*, pp. 136–9; J.M. Wood, *Company Housing for the Shropshire Iron Foundries*, University of Liverpool dissertation for Royal Institute of British Architects, 1974; W. Grant Muter, *The Buildings of an Industrial Community*, London, Phillimore, 1979, pp. 36–44.

46 Minute Books of the Coalbrookdale Company, 1789–97, IGMT archives, CBD 59.82.3.

47 Wood, op. cit.

48 Anon., *A Description of Coalbrookdale in 1801*, IGMT, 1979.

49 Catherine Clark and Judith Alfrey, *Coalport and Blists Hill: Second Interim Report of the Nuffield Survey*, Ironbridge, Ironbridge Institute, 1986, pp. 89–103.

50 SRO 1681/181 and 1987/31/5.

51 The development of Jug Row appears to have been begun by one of the operating companies of the China Works, Horton, Pugh & Anstice, as tenants to Francis Darby. Eight houses were shown on a lease map of 1805 – there were 14 dwellings by 1808 (SBL 18992 and IGMT 1976.103).

52 In 'Workers' housing in the cotton factory colonies', Chapman links the continued use of existing housing in early textile communities with capital shortages in the industry; the reluctance of entrepreneurs to engage in house-building in other than exceptional circumstances is discussed briefly in Rule, op. cit., pp. 99–101.

53 Jeremy Lowe, 'Housing as a source for industrial history', *Industrial Archaeology*, Vol. 8, 1982, pp. 13–35; Chapman, op. cit.

54 This point is discussed in connection with the organisation of the ironworks in Clark and Alfrey, *Coalbrookdale*, pp. 114–15 and 118; a similar pattern of development in those areas of the Gorge associated with mining and other manufactures suggests a similar social structure: the use of capital derived from coal-mining and barge-owning in building is discussed above, chapter 3.

55 Neither in Coalbrookdale nor in Coalport were any community or public buildings provided by the developer.

56 M. Beresford, 'The back-to-back house in Leeds, 1787–1937', in S.D. Chapman, ed, *A History of Working-Class Housing*, Newton Abbot, David & Charles, 1971; Robert Newton, 'Exeter 1770–1870', in M.A. Simpson and T.H. Lloyd, eds, *Middle-Class Housing in Britain*, Newton Abbot, David & Charles, 1977.

57 C.W. Chalklin, *The Provincial Towns of Georgian England: A Study of the Building Process*, London, Edward Arnold, 1974, chapter 5; Judith Alfrey, 'The poor houses of the Lanes: working-class housing in Carlisle', M.A. dissertation, University of Manchester, 1985; Richard Dennis, op. cit., pp. 151–3.

58 Chalklin, op. cit., pp. 57–9, 167–8.

59 Beresford, 'The back-to-back house in Leeds', S.D. Chapman, 'Working-class housing in Nottingham during the Industrial Revolution', in S.D. Chapman, ed, *The History of Working-Class Housing*, Newton Abbot, David & Charles, 1971.

60 Chalklin, op. cit.

61 Development in Paradise, Coalbrookdale, involved the subdivision of formerly larger fields as building plots; it is recorded in a lease map of 1827 (an unreferenced dye-line in IGMT library), and the Tithe Map of 1847.

62 The development of villas in large plots is discussed in Richard Roger, 'Rents and ground rents: housing and the land market in nineteenth century Britain', *Journal of Historical*

Geography, vol. 4, 1979, pp. 129–44; for urban fringe belts, see M.R.G. Conzen, 'The morphology of English towns during the industrial era', in J.W.R. Whitehand, ed., *The Urban Landscape*, London, Institute of British Geographers Special Publication no. 13, 1981.

63 This information is taken from the Tithe Apportionments for the parishes of Madeley and Broseley.

64 In 1829, for example, a lease of land in Coalbrookdale specified an annual rent set at £10 with a supplement for all buildings after the first – leases of other plots of land in Coalbrookdale were quickly followed by building. In Ironbridge, leasing was also common practice on the blocks of land held by Reynolds, and one lease of a malthouse specified a rent of £32 a year, with an additional 5 shillings for every dwelling house which might thereafter be built (SRO 1987/11, 12, 21, and SRO 1681 box 143).

65 Examples of this are 12–13 Tontine Hill, which was designed as a pair, and 25–27 High Street, which is a single building incorporating three separate units.

66 The contrast between the pattern of development, and the homogeneity of building style, characterised settlement in the Gorge from the early eighteenth century. The major models of practice for building are discussed below, in chapter 8. In 'Rents and ground rents', Richard Roger links the need for economy in building working-class housing to the homogeneity of design, contrasting this with the more flamboyant designs commissioned by the middle classes.

67 Title deeds reveal the extent of leasing which is also documented in the Tithe Apportionments.

68 Title deeds for the Ironbridge area after *c.*1790 mention a carpenter, a schoolmaster, two merchants, two maltsters, a draper, a shoemaker, barge-owners and gentry. This contrasts with deeds from an earlier period in which occupations are more commonly miners and colliers, bargemen, foundrymen and labourers. A similar pattern is revealed in Coalbrookdale, where carpenters, colliers, iron-moulders and smiths dominate transactions before *c.*1790: thereafter, cordwainers, victuallers, a tea-dealer, and a banker are listed. After *c.*1803, however, property was increasingly concentrated in the hands of the Darby family. In Broseley, early deeds record colliers, carpenters, a bricklayer among yeomen and gentry – but the Forresters were also acquiring property on a large scale after *c.*1795 (SRO 1224, 1681 and 1987).

69 Where new building was mismatched with the demand for housing, larger houses intended for a middle class were sometimes subdivided as tenements: in the Gorge, new building seems to have found an appropriate market. Although some of the houses built in the late eighteenth to early nineteenth century were quickly altered by the addition of further dwellings, there are very few examples of subdivision; the effects of an imperfect housing market are discussed in M.J. Daunton, *Coal Metropolis: Cardiff 1870–1914*, Leicester, Leicester University Press, 1977, chapter 5.

70 Census analysis reveals that, in 1851, the highest proportion of professional, craft and retail occupations was in that part of Ironbridge between Lincoln Hill, The Wharfage/High Street and Church Hill: the highest concentration of labourers, miners and ironworkers was in the area to the east of Madeley Bank. Overcrowding was even more severe in 1861, but this was largely due to an influx of labour associated with construction of the Severn Valley Railway.

71 Colin Pooley, 'Choice and constraint in the nineteenth-century city: a basis for residential segregation', in J.H. Johnson and Colin Pooley, eds, *The Structure of Nineteenth-Century Cities*, London, Croom Helm, 1982.

72 H.J. Dyos and D.A. Reeder, 'Slums and suburbs', in H.J. Dyos and Michael Wolff, eds, *The Victorian City: Images and Realities*, vol. 2, London, Routledge & Kegan Paul, 1983, pp. 359–86.

73 42–43 Bridge Road, Benthall, and 2–5 Severn Terrace, Ironbridge, are examples of the incorporation of early buildings into an early-nineteenth-century terrace; 22–25 Church Hill, 24 St Luke's Road and Severn Bank, Ironbridge, incorporate earlier building fabric; the plan of The Chestnuts, Coalbrookdale, was constrained by two earlier buildings which were on the site.

74 These aspects of public housing provision have been fully researched elsewhere: see, for example, Ian Bentley, Ian Davis and Paul Oliver, *Dunroamin: The Suburban Semi and Its Enemies*, London, Barrie & Jenkins, 1981.

75 A number of houses were built following the demolition of earlier buildings – see, for example, 27–28 The Wharfage, Ironbridge, 10–12 King Street, Broseley; bungalows at The Knowle, Jackfield, are said to have been built by former occupiers of older property nearer the river.

76 The current revival of interest in surviving older properties and their modernisation and adaptation to new uses is a complex process in which the intervention of a public authority through planning and building forms a part. The impact of regulation, industrial decline, and revival under the aegis of Telford New Town on the evaluation of the building stock, and their implications for the design of new property will form part of the subject of Alfrey, *Living with History* (forthcoming).

8 The making of an industrial vernacular

1 See above, Industrial Foundations, and Landscape of Housing: the Geography of Settlement.

2 Samuel Parsons' Map of Broseley, SRO 1224/1/32; Map of Langley's Tenement, SRO 1224/1/33.

3 To the south of Priory Common, in Woodlands Green and 'Rotherhurst's Tenement', buildings survive on a number of the plots shown on these maps. None is earlier than the first half of the eighteenth century. In the area of Woodlands Green, a number of other buildings, which were shown on the Tithe Map and which may have been early survivals, were demolished later in the nineteenth century.

4 There is, for example, a building on Queen Street where a timber frame has been encased in a later brick core – a star-shaped brick stack may belong to the same building phase as the framing and is probably early seventeenth century; traces of timber frame also survive in buildings on Fox Lane.

5 W.G. Hoskins, 'The re-building of rural England, 1570–1640', *Past and Present*, vol. 4, 1953, pp. 44–59, reprinted in *Provincial England*, 1963; R. Machin, 'The great re-building: a re-assessment', *Past and Present*, vol. 28, 1977, pp. 33–56; Eric Mercer, *English Vernacular Houses*, London, HMSO, 1975, pp. 8–9.

6 Stuart Wrathmell, 'The vernacular threshold of northern peasant houses', *Vernacular Architecture*, vol. 15, 1984, pp. 29–33.

7 One building of this type has been investigated archaeologically, and is discussed in IGMAU BE 84; other buildings of this kind are discussed in Clark and Alfrey, *Benthall and Broseley Wood*, pp. 103–4.

8 These buildings are discussed in Catherine Clark and Judith Alfrey, *Coalbrookdale: First Interim Report of the Nuffield Survey*, Ironbridge, Ironbridge Institute, 1986; *Benthall and Broseley Wood: Third Interim Report of the Nuffield Survey*, Ironbridge Institute, 1987; and *Ironbridge: Fifth Interim Report of the Nuffield Survey*, Ironbridge, Ironbridge Institute, forthcoming; Lloyds Hall is discussed and illustrated in H.E. Forrest, *The Old Houses of Wenlock*, Shrewsbury, Wilding & Son, 1922, p. 89.

9 Among the larger houses of this phase were Lloyds Hall and the house which later became the Dog and Duck: they belong to a rural building tradition and are notably less sophisticated than contemporary building in such towns as Bridgnorth, Much Wenlock and Shrewsbury; a single-unit timbered cottage survived on Jockey Bank, Ironbridge, until recently – a similar cottage survives on Coalport Road, Broseley; 34–35 Belmont Road, Ironbridge, is now two dwellings but may have been built as a single dwelling with a two–unit-plan form; other buildings of this phase are detailed in the interim reports of the Nuffield Survey.

10 Herefordshire Record Office, Diocesan Probate Records.

11 The use of rural forms in early industrial housing is paralleled elsewhere: Jeremy Lowe, *Welsh Industrial Workers' Housing*, Cardiff, National Museum of Wales, 1977; Christian Devilliers and Bernard Huet, *Le Creusot: naissance et développement d'une ville industrielle 1782–1914*, Macon, Champ Vallon, 1981, pp. 164–6.

12 The two surviving buildings are built on waste land above Wesley Road, Ironbridge.

13 Royal Commission on Historical Monuments, *Workers Housing in West Yorkshire, 1750–1920*, London, HMSO, 1986, chapter 1; Barrie Trinder, *The Industrial Revolution in Shropshire*, London, Phillimore, 1981, p. 188.

14 A few buildings of this type survive in Lightmoor; a similar building, restored to its original nineteenth-century form can be seen at the Blists Hill Open Air Museum; for the poverty of rural dwellings, see J.H. Bettey, 'Seventeenth century squatter dwellings: some documentary evidence', *Vernacular Architecture*, vol. 13, 1982, pp. 28–30; Enid Gauldie, *Cruel Habitations, A History of Working Class Housing 1780–1918*, London, George Allen & Unwin, 1974, pp. 22–4; John Burnett, *A Social History of Housing 1815–1970*, London, Methuen, 1980, pp. 30–47.

15 R.W. Brunskill, *Illustrated Handbook of Vernacular Architecture*, London, Faber & Faber, 1974, pp. 26–7; RCHM, *Rural Houses of the Lancashire Pennines 1560–1760*, London, HMSO 1986, pp. 81–84.

16 Machin, op. cit.; Wrathmell, op. cit.; R.J. Brown, *The English Country Cottage*, London, Hamlyn, 1979, p. 20.

17 Raddle Hall, Broseley, is dated 1663; Bedlam Hall, Madeley Wood, was described as a brick Jacobean mansion in Forrest, op. cit., p. 100, but as timber-framed in John Randall, *A History of Madeley*, Madeley, 1880, p. 333.

18 24 Hodgebower and 9 Chapel Road, Ironbridge, 40–41 Quarry Road, Broseley Wood; title deeds for Tea Kettle Row are in SRO 1681/179.

19 De Villiers and Huet, op. cit., pp. 154–60; Henry Glassie, *Folk Housing in Middle Virginia*, Knoxville, Tenn., University of Tennessee Press, 1975, chapter 6; Thomas Hubka, 'Just folks designing: vernacular designers and the generation of form', in Dell Upton and John Michael Vlach, eds, *Common Places: Readings in American Vernacular Architecture*, Athens, Ga, University of Georgia Press, 1986, pp. 426–33.

20 Glassie, op. cit., chapter 4; Glassie, 'Eighteenth-century cultural process in Delaware Valley folk building', Hubka, 'Just folks designing', Dell Upton, 'Vernacular domestic architecture in eighteenth-century Virginia', in Upton and Vlach, eds, op. cit., pp. 394–425, 426–32, 315–35.

21 See, for example, 34 Hodgebower, 34–35 Belmont Road, Ironbridge, 42–43 Bridge Road, Benthall, 57–59 Hodgebower, Ironbridge (demolished); a number of stone and timber or brick and timber buildings now demolished in Madeley Wood and Jockey Bank are documented in photographs in the IGMT archive.

22 While the exact location of original openings in timber-framed cottages cannot always be determined, the arrangement of wide dormers in Yew Tree Cottage, Coalbrookdale, and 68 Ladywood serve as examples of a form which was not replicated in later building, and contrasts with the simpler construction used in smaller cottages in Madeley Wood.

23 See, for example, cottages at Severnside (demolished), 12 Chapel Road, Ironbridge; G.D. Newton, 'Single-storey cottages in West Yorkshire', *Folk Life*, vol. 14, 1976, pp. 65–74.

24 See, for example 14–15 Woodlands Road, 57–59 Hodgebower, Ironbridge (demolished); a timber-framed cottage on Coalport Road, Broseley, has a doorway against the stack – this was also the original arrangement in a stone cottage at Darnley, near Broseley, where the doorway was later moved to the centre of the façade when the building was refronted.

25 Herefordshire Record Office, Diocesan Probate Records.

26 One will suggests an arrangement of this sort, where a house had been divided into three (John Bayley, collier, 1712); structural survey of 22 The Wharfage, Ironbridge, during alterations carried out by IGMAU revealed that this building had also gone through a similar sequence of development.

27 The close correspondence of house size and social rank or economic means is taken as an axiom of vernacular architecture: see, for example, Brunskill, op. cit., p. 22.

28 Springhill, Woodside; another house of this type is illustrated in Vivares' engraving of the Upper Works, and reproduced in Stuart Smith, *A View from the Iron Bridge*, IGMT, 1979, p. 14; other buildings with a similar form appear in illustrations of the riverside at Benthall, Loadcroft Wharf and the Lloyds (IGMT).

29 The eighteenth-century houses at Coalford, Jackfield, range from a single-unit cottage to a large two-unit house, but all employ the same basic vocabulary; a similar range of variation is apparent in Broseley Wood.

30 See, for example, 4 Speeds Lane, Broseley, where asymmetry is also associated with poorer standards of construction and decoration – wood lintels instead of cambered brick window heads, and an absence of the ornamented eaves band which characterises many houses of this period.

31 Probate records; Hearth Tax returns of 1672 also illustrate this distribution – for example, of 28 houses in Benthall, 13 had two hearths and 5 had one hearth: *The Shropshire Hearth Tax Return of 1672*, Shropshire Archaeological and Parish Register Society, 1949; Smith, op. cit., p. 14.

32 Glassie, *Folk Housing in Middle Virginia*, chapters 4 and 6.

33 Brunskill, 'Traditional domestic architecture of south-west Lancashire', *Folk Life*, vol. 15, 1977; Newton, op. cit.

34 Brunskill, *Houses*; M.W. Barley, *The English Farmhouse and Cottage*, London, Routledge & Kegan Paul, 1961; the convention of classifying buildings according to plan types and arranging these in a series beginning with a single unit as the most basic and therefore typologically earliest form may overlook the fact that the single-unit cottage became more widely used: W. Grant Muter, *The Buildings of an Industrial Community*, London, Phillimore, 1979, pp. 28–30.

35 A number of single-unit cottages with only a tiny additional service room were built in the second half of the nineteenth century: see, for example, 8–16 Church Road, Coalbrookdale.

36 See, for example, Springhill, Woodside, Coalbrookdale; Calcutts House, Jackfield; 11 The Wharfage, Ironbridge.

37 For examples of buildings designed as pairs, see 40–41 Simpsons Lane, Broseley Wood, 38–39 Hodgebower, Ironbridge.

38 For examples of the terrace as a row of independent dwelling-units, see 8–14 Barratts Hill, Broseley, The Calcutts, Jackfield, Panshop Bank, Paradise, Coalbrookdale (demolished).

39 See, for example, 15–17 and 31–35 Hodgebower, Ironbridge.

40 K. Jones, *et al.*, 'Holywell Lane: a squatter community in the Shropshire Coalfield', *Industrial Archaeology Review*, vol. 6, 1982, pp. 163–85.

41 Mercer, op. cit., p. 133.

42 Compare, for example, 22 and 27 Cobwell Road with 4 Simpsons Lane, Broseley; decorative brickwork is used on the chimneys of 24 Hodgebower, dated 1714; decorative Flemish Bond is used on 36 and 44–47 Wellington Road, Coalbrookdale.

43 Evidence for early builders is, however, sparse: probate records only record one glazier, and one stonemason and bricklayer during the period *c.*1660–1760.

44 Herefordshire Record Office, Diocesan Probate Records, John Brooke, 27 March 1721.

45 Royal Commission on Historical Monuments, *Workers' Housing in West Yorkshire*, op. cit., chapters 2 and 3; J. Lowe, 'Housing as a source for industrial history', *Industrial Archaeology*, vol. 8, 1982, pp. 13–35; Barley, op. cit.

46 Charity Row, Carpenter's Row, Chapel Row and Engine Row were all built by the Coalbrookdale Company *c.*1781–9: Clark and Alfrey, *Coalbrookdale*, pp. 130–4, 136–9.

47 See, for example, The Cottage and Jasmine Cottage, Cherry Tree Hill, Coalbrookdale.

48 'Map referring to the Annexed Lease', 1827, unreferenced dye-line, IGMT liabrary; Tithe Map, 1847.

49 Information made available by owner from privately held title deeds.

50 14 Wellington Road.

51 Three houses on Cherry Tree Hill employ this detail – they were built by 1851 (Plans for the Great Western Railway, SRO 1969); houses with similar detailing in Ironbridge all postdate the Tithe Survey of 1847.

52 This contrasts with housing in Madeley, for example, where a very similar model is used employing a central window over the door, sometimes blind.

53 These two houses, built to adjoin each other, employ an identical style with small variations in size – they were not built as a pair, however – only one existed in *c.*1860, the date of a photograph.

54 Evidence for the extent and organisation of building work is taken from the 1851 census; additional, but less comprehensive information is contained in the trade directories for the period.

55 Cast-iron lintels for window heads were commonplace, though very few houses used cast-iron for the windows themselves. A range of decorative brick was available from *c.*1830, and white brick was especially popular in the growth of Ironbridge. Blue, red, white and 'kiss-marked' bricks were all exploited for their decorative properties: The Orchard, Church Hill, Ironbridge, built for a Jackfield brickmaker, is a text-book for this variety, but many smaller houses were built with decorative string courses, or a sharply contrasting Flemish Bond (for example Raymore, Paradise, Coalbrookdale).

56 A number of these houses were built with a definite orientation to a street-line, apparent on Wellington Road, Coalbrookdale, Hodgebower and Madeley Bank, Ironbridge; the parallel formation of essentially urban facilities is documented in Muter, op. cit., pp. 10–22.

57 Quoted in I. Lawley, 'Art and ornament in iron: design and the Coalbrookdale Company', 1979, MS in IGMT library.

58 Quoted in Trinder, *The Industrial Revolution in Shropshire*, p. 199. These houses were not built by the Coalbrookdale Company, though their work elsewhere at Horsehay and Sandybank Row also makes a distinction between housing for the 'superior artisan' and the general labourer, clearly marked in the degree of architectural sophistication.

59 Arthur Raistrick, *Dynasty of Ironfounders*, London, Longman, Green & Co., 1953, pp. 256–63.

60 In 1851, for example, Paradise House was the home of Charles Crookes, then an assistant to an ironfounder, later listed as an ironmaster himself. His household included two servants. Numbers 11 and 12 Paradise were occupied by a modeller and an accountant, and Prospect House on Church Road was occupied by a draughtsman at the iron foundry.

61 The Hawthorns and 57 Belmont Road.

62 J. Randall, op. cit., pp. 346–50.

63 The building of the police station is documented in the Minute Books of the Madeley Sanitary Committee, Wenlock Borough Records, 1859–63.

64 See, for example, 29 Church Hill and 25–30 Hodgebower, Ironbridge.

65 15–22 Woodside.

66 Minute Books of the Madeley Sanitary Committee.

67 Tudor Walters Report, 1918.

68 Minute Books of the Madeley Sanitary Committee.

69 Ibid.

70 The desire for uniformity as an expression of community, as well as of economy, did not necessarily lead to adoption of the neo-Georgian style: Ian Bentley, 'Individualism or community? Private enterprise housing and the council estate', in Ian Bentley, Ian Davis and Paul Oliver, *Dunroamin: The Suburban Semi and Its Enemies*, London, Barrie & Jenkins, 1981, pp. 104–21.

71 Minute Books of the Madeley Sanitary Committee.

72 9–19 Cobwell Road, Broseley Wood, Fernlea, Riversdale, etc. Bridge Road, Benthall;

there are also a number of bungalows in Jackfield where a basic model has been adapted to different specifications – see Calcutts Road and the Knowle, for example; the interwar development of Chapel Road, Jackfield, used an identical plan for 10 dwellings: small differences occur in the detailing.

73 5–6 Ladywood, Ironbridge, 41–42 Wellington Road, Coalbrookdale.

74 See for example, 'Au Pont de Fer', High Street Ironbridge; Beddows Building, Madeley Bank, Ironbridge. The social dimensions of purpose-built shops like these are discussed in Leonore Davidoff and Catherine Hall, *Family Fortunes: Men and Women of the English Middle Class 1780–1850*, London, Hutchinson, 1987, p. 366.

75 Former shop adjacent to Jackfield Church built largely from corrugated-iron sheet and reused components.

9 The industrial landscape today: from analysis to protection

1 UNESCO, *Convention for the Protection of the World Cultural and National Heritage.*

BIBLIOGRAPHY

Alfrey, J., 'The poor houses of the Lanes: working-class housing in Carlisle', University of Manchester, M.A. dissertation, 1982.

—— *Living with History*, forthcoming.

Allied Ironfounders Ltd, *Allied Ironfounders in Shropshire*, Ketley, Allied Ironfounders, 1959.

Anon., 'The Coalbrookdale tramroad', *The Locomotive Magazine*, 15 December 1917, pp. 443–45.

Anon., 'Excavations at Lloyds Coppice Engine, Ironbridge', unpublished MS, IGMT library, 1970.

Architects' Compendium and Catalogue, 1901.

Aston, M., *Interpreting the Landscape*, London, Batsford, 1979.

Atkinson, D., *Tobacco Pipes of Broseley, Shropshire*, London, privately printed, 1970.

Baker, D.W., *Swannington Festival Group Development Proposals*, privately printed, 1984.

Barker, P., *Techniques of Archaeological Excavation*, London, Batsford, 1982.

Barley, M.W., *The English Farmhouse and Cottage*, London, Routledge & Kegan Paul, 1961.

Barnard, J., *Victorian Ceramic Tiles*, London, Studio Vista, 1972.

Beale, R., *The Old Wind: A Preliminary Report*, Ironbridge, IGMAU, 1988.

Beaver, S., 'The Potteries: a study in the evolution of a cultural landscape', *Institute of British Geographers Transactions and Papers*, 1964, vol. 4, pp. 1–31.

Benthall, E., 'Some xviiith century Shropshire potteries', *Transactions of the Shropshire Archaeological Society*, 1957, vol. 55, pp. 159–70.

Bentley, I., 'Individualism or community? Private enterprise housing and the council estate', in I. Bentley, I. Davis and P. Oliver, *Dunroamin: The Suburban Semi and its Enemies*, London, Barrie & Jenkins, 1981.

Beresford, M.W., 'The back-to-back house in Leeds, 1787–1937', in S.D. Chapman, ed., *The History of Working-Class Housing: A Symposium*, Newton Abbot, David & Charles, 1971.

Beresford, M. W. and Hurst, J.G., eds, *Deserted Medieval Villages*, Woking, Lutterworth Press, 1971.

Beresford, M.W., Hurst, J.G. and Rahtz, P.A., *Wharram: A Study of Settlement on the Yorkshire Wolds*, London, Society for Medieval Archaeology, 1987.

Bettey, J.H., 'Seventeenth-century squatter dwellings: some documentary evidence', *Vernacular Architecture*, 1982, vol. 13, pp. 28–30.

Biddle, M., 'Metres, areas and robbing', *World Archaeology*, 1969, vol. 1, pp. 208–19.

Blackman, J., 'The development of the retail grocery trade in the nineteenth century', *Business History*, 1962, vol. 5.

Blackwall, A., *Historic Bridges of Shropshire*. Shrewsbury, Shropshire Libraries & the Highways and Transport Department, 1985.

Blake Roberts, D. and Blake Roberts, G., 'The results of recent excavations in Coalport, Shropshire', *English Ceramic Circle Transactions*, 1981, vol. 2, pp. 71–80.

Blanchard, I., 'Miners and the agricultural community', *Agricultural History Review*, 1970, vol. 20, pp. 96–106.

Braudel, F., *Civilisation and Capitalism Fifteenth to Eighteenth Century*, vol. 2, *The Wheels of Commerce*, London, Collins, 1983.

Broadbridge, S.R., 'Joseph Banks and the West Midlands iron industry', *Staffordshire Industrial Archaeology Society Journal*, 1971, vol. 2, pp. 1–20.

Brown, I., 'The mineral wealth of Coalbrookdale', *Bulletin of the Peak District Mines Historical Society*, 1965, vol. 2, pp. 255–89.

—— 'The Tar Tunnel, Coalport', *Shropshire Newsletter*, 1967, vol. 32.

—— 'Mineral working and land reclamation in the Coalbrookdale Coalfield', Ph.D. thesis, Leicester University, 1975.

—— 'Underground in the Ironbridge Gorge', *Industrial Archaeology Review*, 1979, vol. 3, pp. 158–69.

—- 'Some mining accidents in the Broseley Field', *Journal of the Wilkinson Society*, 1980, vol. 8, pp. 3–5.

Brown, I.J. and Trinder, B.S., *The Tar Tunnel*, Ironbridge, IGMT, 1979.

Brown, R.J., *The English Country Cottage*, London, Hamlyn, 1979.

Brunskill, R.W., 'Traditional domestic architecture of south-west Lancashire', *Folk Life*, 1977, vol. 15, pp. 65–80.

—— *Illustrated Handbook of Vernacular Architecture*, London, Faber & Faber, 1978.

—— *Traditional Buildings of Britain*, London, Gollancz, 1981.

Buist, W., 'A survey of the Engineering Shop, one of the buildings belonging to the Coalbrookdale Company', unpublished assignment, Ironbridge Institute, University of Birmingham, 1991.

Burnett, J., *A Social History of Housing 1815–1970*, London, Methuen, 1980.

Carter, H., 'Towns and urban systems', in R.A. Dodgson and R.A. Butlin, eds, *A New Historical Geography of England and Wales*, London, Academic Press, 1978.

Carter, H. and Wheatley, J.S., 'Some aspects of the spatial structure of two Glamorgan towns in the nineteenth century', *Welsh History Review*, 1978–9, vol. 9, pp. 32–56.

Celoria, F. and Kelly, J., 'A post-medieval pottery site and a kiln base found off Albion Square, Hanley, Stoke on Trent, Staffordshire, England SJ 885474', *City of Stoke on Trent Museum Archaeological Society Report*, 1973, vol. 4.

Chaffers, W., *The New Chaffers Marks and Monograms on Pottery and Porcelain*, ed. Frederick Litchfield, 13th edn, London, Reeves & Turner, 1912.

Chalklin, C.W., *The Provincial Towns of Georgian England, A Study of the Building Process 1740–1850*, London, Edward Arnold, 1974.

Chambers, J.D. and Mingay, G.E., *The Agricultural Revolution, 1750–1800*, London, Batsford, 1966.

Chapman, S.D., 'Working-class housing in Nottingham during the Industrial Revolution', in S.D. Chapman, ed., *The History of Working Class Housing: A Symposium*, Newton Abbot, David & Charles, 1971.

—— 'Workers' housing in the cotton factory colonies 1770–1850', *Textile History*, 1976, vol. 7, pp. 112–39.

—— *The Cotton Industry in the Industrial Revolution*, London, Macmillan, 1987.

Children's Employment Commission, *Children's Employment Commission Report*, British Parliamentary Papers 16, report no. 381, 1842.

—— *Children's Employment Commission Fifth Report*, British Parliamentary Papers, 1862.

Clark, B.D. and Gleave, M.B., eds, *Social Patterns in Cities*, Institute of British Geographers Special Publication 5, London, Institute of British Geographers, 1973.

Clark, C., *Horsehay Ironworks, Shropshire*, Report for Wrekin District Council, Ironbridge, Ironbridge Institute, 1989.

—— *Archaeological and Historical Issues Relating to the Proposed New Road Bridge at Ladywood, Ironbridge*, report prepared for English Heritage, Ironbridge, Ironbridge Institute, 1989.

—— *Ironbridge Gorge*, London, Batsford, 1993.

Clark, C. and Alfrey, J., *Coalbrookdale: First Interim Report of the Nuffield Survey*, Ironbridge: Ironbridge Institute, 1986.

—— *Coalport and Blists Hill: Second Interim Report of the Nuffield Survey*, Ironbridge, Ironbridge Institute, 1986.

—— *Benthall and Broseley Wood: Third Interim Report of the Nuffield Survey*, Ironbridge, Ironbridge Institute, 1987.

—— *Jackfield and Broseley: Fourth Interim Report of the Nuffield Survey*, Ironbridge, Ironbridge Institute, 1989.

—— *Ironbridge: Fifth Interim Report of the Nuffield Survey*, Ironbridge, Ironbridge Institute, forthcoming.

Clark, C. et al., *Caughley Saggar Works: An Archaeological Evaluation*, Ironbridge, Ironbridge Institute, 1988.

Clark, C., Horton M. *et al.*, *Duncote Farm, Atcham: An Archaeological Evaluation in Advance of the A5 By-pass*, Ironbridge, Ironbridge Institute, 1989.

Clark, C. and Winkworth, A., *Duncote Farm, Atcham*, for English Heritage, Ironbridge, Ironbridge Institute, 1990.

Clark, C., Winkworth, A. and Worthington, M., *Ladywood Bridge Site, Ironbridge, Telford: Excavation and Further Documentary Research*, Report for Shropshire County Council, Ironbridge, Ironbridge Institute, 1990.

Clarke, D., *Analytical Archaeology*, 2nd edn, London, Methuen, 1978.

Clarkson, L.A., *Proto-industrialisation: The First Phase of Industrialisation*, Basingstoke, Macmillan, 1985.

Clifton-Taylor, A., *The Pattern of English Building*, London, Faber, 1972.

Collins, P., *Walsall Leather Industry*, Ironbridge, Ironbridge Institute, 1989.

Conzen, M.R.G., 'The growth and character of Whitby', in G.H.J. Daysh, ed., *A Survey of Whitby and the Surrounding Area*, Eton, 1958.

—— *Alnwick, Northumberland: A Study in Town-plan Analysis*, Institute of British Geographers Special Publication no. 27, London, Institute of British Geographers, 1960.

—— 'The morphology of towns in Britain during the industrial era', in J.W.R. Whitehand, ed., *The Urban Landscape: Historical Development and Management*, Institute of British Geographers, Special Publication no. 13, London, Institute of British Geographers, 1981.

Corfield, P.J., *The Impact of English Towns, 1700–1800*, Oxford, Oxford University Press, 1982.

Cossons, N. and Trinder, B., *The Iron Bridge*, Bradford-on-Avon, Moonraker Press and IGMT, 1979.

Court, W.H.B., *The Rise of the Midland Industries 1600–1838*, London, Oxford University Press, 1938.

Cox, N., 'Imagination and innocation of an industrial pioneer: the first Abraham Darby', *Industrial Archaeology Review*, 1990, vol. 12, pp. 127–44.

Crosfield, M.C. and Johnstone, M.S., 'A study of Ballstone and the associated beds', *Proceedings of the Geological Association*, 1914, vol. 25, pp. 207–9.

Crossley, D. and Cleere, H., *The Iron Industry of the Weald*, Leicester, Leicester University Press, 1985.

Daniel, P. and Hopkinson, M., *The Geography of Settlement*, Edinburgh, Oliver & Boyd, 1989.

Darley, G., *Villages of Vision*, London, Paladin, 1978.

Daunton, M., *Coal Metropolis: Cardiff 1870–1940*, Leicester, Leicester University Press, 1977.

—— *House and Home in the Victorian City: Working Class Housing 1850–1914*, London, Edward Arnold, 1983.

Daunton, M., 'Public place and private space', in D. Fraser, and A. Sutcliffe, eds, *The Pursuit of Urban History*, London, Edward Arnold, 1983.

Davidoff, L. and Hall, C., *Family Fortunes: Men and Women of the English Middle Class 1780–1850*, London, Hutchinson, 1987.

Dawes, N.M., 'History of brick and tile production on the Coalbrookdale Coalfield', unpublished MS, IGMT library, 1979.

Defoe, D., *A Tour through the Whole Island of Great Britain*, first published 1724–6, Harmondsworth, Penguin, 1979.

Dennis, R., 'Stability and change in urban communities: a geographical perspective', in J.H. Johnson, and C. Pooley, eds, *The Structure of Nineteenth Century Cities*, London, Croom Helm, 1962.

—— *English Industrial Cities of the Nineteenth Century: A Social Geography*, Cambridge, Cambridge University Press, 1984.

Dennis, R. and Clout, H., *A Social Geography of England and Wales*, London, Pergamon, 1980.

Devilliers, C. and Huet, B., *Le Creusot: naissance et développement d'une ville industrielle 1782–1914*, Macon, Champ Vallon, 1981.

Dobson, E., 'A rudimentary treatise on the manufacture of bricks and tiles', reprinted in F. Celoria, ed., *Journal of Ceramic History*, 1971, vol. 5.

Dodgson, R.A. and Butlin, R.A., eds, *A New Historical Geography of England*, London, Academic Press, 1978.

Drury, P.J., 'Post-medieval brick and tile kilns at Runsell Green, Danbury, Essex', *Post-Medieval Archaeology*, 1975, vol. 9, pp. 203–21.

Duckworth, S., 'Benthall to Coalport: an interpretive assessment of the south bank of the River Severn in the Ironbridge Gorge', dissertation, Ironbridge Institute, University of Birmingham, 1988.

Dyos, H.J. and Reeder, D.A., 'Slums and suburbs', in H.J. Dyos, and M. Wolff, *The Victorian City: Images and Realities*, vol. 2, London, Routledge & Kegan Paul, 1983.

Edmundson, R.S., 'The Coalport China Works: a study of the circumstances and dating of the riverside group of buildings, particularly the "Smuggler" area', unpublished MS, IGMT library, n.d.

—— 'Coalport China Works, Shropshire: a comparative study of the premises and the background to their study', *Industrial Archaeology Review*, 1979, vol. 3, pp. 122–45.

—— 'Bradley & Co., Coalport Pottery 1796–1800', *Transactions of the Northern Ceramic Society*, 1981, vol. 4, pp. 127–55.

Eele, M. 'An account of making Pitch, Tar and Oil out of a blackish stone in Shropshire, communicated by Mr Martin Eele, the Inventor of it', *Royal Society Philosophical Transactions*, 1697, vol. 19, p. 544.

Falconer, K., *Guide to England's Industrial Heritage*, London, Batsford, 1980.

Finberg, H.P.R., *Early Charters of the West Midlands*, 2nd edn, Leicester, Leicester University Press, 1971.

Firman, R.J. and Firman, P.E., 'A geological approach to the study of medieval bricks', *The Mercian Geologist*, 1967, vol. 2, pp. 299–317.

Fisher, F.J., *Essays in the Economic and Social History of Tudor and Stuart England in Honour of R.H. Tawney*, Cambridge, Cambridge University Press, 1961.

Forrest, H.W., *The Old Houses of Wenlock*, Shrewsbury, Wilding & Sons, 1915.

Fraser, D. and Sutcliffe, A., eds, *The Pursuit of Urban History*, London, Edward Arnold, 1983.

Frost, P., 'Yeomen and metalsmiths: livestock in the dual economy in south Staffordshire, 1560–1720', *Agricultural History Review*, 1981, vol. 29, pp. 29–41.

Gameson, F.R., 'An 8-man pit – an historic beam engine and the craft of Harry Roberts', *Shropshire Magazine*, 20–1, March 1952.

Gaskell, S.M., *Building Control, National Legislation and the Introduction of Local Bye Laws in Victorian England*, London, published for the British Association for Local History by Bedford Square Press, 1983.

Gauldie, E., *Cruel Habitations, A History of Working Class Housing 1780–1918,* London, George Allen & Unwin, 1974.

Geological Survey and Museum, 'Refractory materials, fireclays', *Special Reports on the Mineral Resources of Great Britain*, 1920, vol. 14, pp. 148–55.

Glassie, H., *Folk Housing in Middle Virginia*, Knoxville, University of Tennessee Press, 1975.

—— 'Eighteenth century cultural process in Delaware Valley folk building', in D. Upton and J.M. Vlach, eds, *Common Places: Readings in American Vernacular Architecture*, Athens, Ga, University of Georgia Press, 1986.

Graslund, B., *The Birth of Prehistoric Chronology: Dating Methods and Dating Systems in Nineteenth-Century Scandinavian Archaeology*, Cambridge, Cambridge University Press, 1981.

Hains, B.A. and Horton, A., *British Regional Geology: Central England*, London, HMSO, 1969.

Hall, J.W., 'Joshua Field's Diary of a tour in 1821 through the Midlands', *Transactions of the Newcomen Society*, 1925–6, vol. 6, pp. 30–2.

Hamblin, R.J.O. *et al.*, 'Mineral resources of the Coalbrookdale Coalfield', *Mercian Geologist*, 1969, vol. 12, pp. 9–27.

Hammond, M., 'The Blists Hill Brickworks Project', unpublished MS, IGMT library, 1979.

Harley, J.B., *Ordnance Survey Maps: A Descriptive Manual*, Southampton, Ordnance Survey, 1975.

Harriss, E.C., *Principles of Archaeological Stratigraphy*, London, Academic Press, 1979.

Henkel, D.J. and Skempton, R.W., 'A landslide at Jackfield, Shropshire in a heavily over-consolidated clay', *Proceedings of the International Conference on the Stability of Earth Slopes*, 1954, pp. 131–7.

Herbert, A.T., 'Jackfield decorative tiles in use', *Industrial Archaeology Review*, 1979, vol. 3, pp. 146–52.

Herbert, D.T., *Urban Geography: A Social Perspective*, Newton Abbot, David & Charles, 1972.

Hey, D., *Rural Metalworkers of the Sheffield Region: A Study of Rural Industry before the Industrial Revolution*, Leicester, Leicester University Press, 1972.

Higgins, D.A., 'Blists Hill mine', unpublished MS, IGMT library, 1985.

—— 'The clay pipe industry of Broseley and Benthall', Ph.D. thesis, University of Liverpool, 1987.

Higgins, D.A. *et al.*, *The Broseley Pipeworks: An Archaeological and Historical Evaluation*, Ironbridge, Ironbridge Institute, 1988.

Holderness, B.A., 'Open and close parishes in England in the eighteenth and nineteenth centuries', *Agricultural History Review*, 1972, vol. 20, pp. 126–39.

Holmes, D.A., 'The working of Wenlock Limestone', Ironbridge Institute, 1987.

Hopwood, A., 'A history of Messrs Maws & Co. 1850–1970', unpublished MS, 1975.

Horton, M.C., *Newdale: An Industrial Township of the Mid-Eighteenth Century*, Ironbridge, Ironbridge Institute, 1987.

—- *Newdale: The North Building*, Ironbridge, Ironbridge Institute, 1989.

Horton, M.C. and Biddle, S., *Granville Country Park: An Evolving Industrial Landscape in Lilleshall, Telford*, Report for Wrekin District Council, Ironbridge, Ironbridge Institute, 1987.

Hoskins, W.G., 'The rebuilding of rural England 1570–1640', *Past and Present*, no. 4, pp. 44–59, reprinted in W.G. Hoskins, *Provincial England*, London, Macmillan, 1953.

—— *The Making of the English Landscape*, Harmondsworth, Penguin, 1955, reprinted 1971.

—— *The Midland Peasant*, London, Macmillan, 1957.

Houghton, A.W.J., 'Caughley Porcelain Works near Brosley, Salop', *Industrial Archaeology Magazine*, 1968, vol. 5, pp. 184–92.

Hubka, T., 'Just folks designing: vernacular designers and the generation of form', in D. Upton and J.M. Vlach, eds, *Common Places: Readings in American Vernacular Architecture*, Athens, Ga, University of Georgia Press, 1986.

Hudson, K., *Food Clothes and Shelter*, London, John Baker, 1978.
Hulbert, C., *History and Description of the County of Shropshire*, Shrewsbury, 1837.
Hutchins, B.L. and Hanson, A., *A History of Factory Legislation*, London, P.S. King & Son, 1903.
Jewitt, L., *Ceramic Art of Great Britain*, London, Virtue, 1878.
Johnson, J.H. and Pooley, C. eds, *The Structure of Nineteenth-Century Cities*, London, Croom Helm, 1962.
Jones, A., *The Coarse Earthenwares,* Ironbridge, IGMAU, 1988.
Jones, E., *Industrial Architecture in Britain 1750–1939*, London, Batsford, 1985.
Jones, E.L., 'The agricultural origins of industry', *Past and Present*, 1968, no. 40, pp. 58–71.
Jones, K., Hunt, M. *et al.*, 'Holywell Lane: a squatter community in the Shropshire Coalfield', *Industrial Archaeology Review*, 1982, vol. 6, pp. 163–85.
Jones, N.W., 'A wooden waggonway at Bedlam Furnace, Ironbridge', *Post-Medieval Archaeology*, 1987, vol. 21, pp. 259–62.
Jones, T., 'Impressions of the Coalbrookdale area', unpublished MS, IGMT library, n.d.
Kain, R.J.P. and Price, H.C., *The Tithe Surveys of England and Wales*, Cambridge, Cambridge University Press, 1985.
Kay, G., 'Minton, Hollins & Co. The tile factory in Shelton Old Road, Stoke-on-Trent', unpublished dissertation, Ironbridge Institute, University of Birmingham, 1988.
Langton, J., *Geographical Change and Industrial Revolution Coal Mining in Southwest Lancashire, 1590–1900,* Cambridge, Cambridge University Press, 1979.
Lawley, I., 'Art and ornament in iron: design and the Coalbrookdale Company', unpublished MS, IGMT library, 1979.
Lawton, R., 'Population and society', in R.A. Dodgson and R.A. Butlin, eds, *A New Historical Geography of England and Wales,* London, Academic Press, 1978.
Leese, J.S., 'Old English power plants', *Power*, 1912, vol. 36, pp. 108–9.
Levine, D., *Family Formation in an Age of Nascent Capitalism*, London, Academic Press, 1977.
Lewis, M.J.T., *Early Wooden Railways*, London, Routledge, 1970.
Lockett, T.A., *Collecting Victorian Tiles*, London, Antique Collectors' Club, 1982.
Lowe, J., *Welsh Industrial Workers' Housing*, Cardiff, National Museum of Wales, 1977.
—- 'Housing as a source for industrial history', *Industrial Archaeology*, 1982, vol. 8, pp. 13–35.
Macdonald, M., 'The Craven Dunnill clay arcs', unpublished essay, Ironbridge Institute, University of Birmingham, 1988.
Machin, R., 'The great re-building, a reassessment', *Past and Present*, 1977, no. 77, pp. 33–56.
—- *The Houses of Yetminster*, Bristol, University of Bristol Department of Extramural Studies, 1978.
Macleod, M., Stratton, M., and Trinder, B., *Llanymynech Hill: An Archaeolgoical and Historical Evaluation*, for L&R Consultants, Ironbridge, Ironbridge Institute, 1987.
Malam, J., 'White salt-glazed stoneware manufactured at Jackfield', *West Midlands Archaeology*, 1981, vol. 24, pp. 45–50.
Marshall, J., *The Severn Valley Railway*, Newton Abbot, David & Charles, 1989.
Mathias, P., *The First Industrial Nation: An Economic History of Britain 1700–1914*, London, Methuen, 1969, reprinted 1974.
Matkin, R.B., 'John Rennie's diary of a journey through northern England, 1784', *East Kent Maritime Trust Historical Study No. 2*, 1986, pp. 23–6.
Maw, G., 'Notes on the drift deposits of the valley of the Severn', *Quarterly Journal of the Geological Society of London*, 1864, vol. 20, p. 142.
Medick, H., 'The proto-industrial family economy: the structural function of household and family during the transition from peasant to industrial capitalism', *Social History*, 1976, vol. 1.
Mendels, F., 'Agriculture and peasant industry in eighteenth-century Flanders', *Journal of Economic History*, 1972, vol. 27, pp. 241–61.

Mercer, E., *English Vernacular Houses*, London, HMSO, 1979.
Messenger, M., *Pottery and Tiles of the Severn Valley*, London, Remploy, 1979.
Mitchell, I., 'The development of urban retailing 1700–1815', in P. Clark, ed., *The Transformation of English Provincial Towns 1600–1800*, London, Hutchinson, 1984.
Mott, R.A., 'Abraham Darby I and II and the coal–iron industry', *Transactions of the Newcomen Society*, 1957, vol. 31, pp. 49–94.
—— 'The Coalbrookdale Group Horsehay Works part I', *Transactions of the Newcomen Society*, 1957, vol. 31, pp. 271–87.
—— 'The Coalbrookdale Group Horsehay Works part II', *Transactions of the Newcomen Society*, 1959, vol. 32, pp. 43–56.
Mugridge, A.J., *Brick and Tile Manufactories in the Severn Gorge*, Jackfield, Orchard Press, 1987.
Murchison, R.I., *Siluria*, 4th edn, London, John Murray, 1867.
Muter, W.G., *The Buildings of an Industrial Community*, Chichester, Phillimore, 1979.
Neff, J.U., *The Rise of the British Coal Industry*, London, Routledge, 1932.
Newton, G.D., 'Single storey cottages in West Yorkshire', *Folk Life*, 1976, vol. 14, pp. 65–74.
Newton, R., 'Exeter 1770–1870', in M.A. Simpson and T.H. Lloyd, eds, *Middle Class Housing in Britain*, Newton Abbot, David & Charles, 1977.
Norris, G., 'A survey of the heavy erecting shop, Coalbrookdale', unpublished essay, Ironbridge Institute, Birmingham University, 1989.
Norris, W., 'Horsehay Ironworks', unpublished MS, IGMT library, 1876.
Oliver, P., Davis, I. and Bentley, I., *Dunroamin: The English Suburban Semi and Its Enemies*, London, Barrie & Jenkins, 1981.
Ove Arup, *Limestone Mines in the Wrekin Area*, Report for Shropshire County Council, 1986.
Pannett, D., 'Fish weirs on the River Severn', *Shropshire Newsletter*, 1972, vol. 41, pp. 4–7.
Parton, T., *Notes on the Shropshire Coalfield*, London, Colliery Guardian Office, 1868.
Pawson, E., 'The framework of industrial change' in R.A. Dodgson, and R.A. Butlin, eds, *A New Historical Geography of England*, London, Academic Press, 1978.
Pears, A., 'Maw and Company: 1900–1910', unpublished dissertation, Ironbridge Institute, University of Birmingham, 1989.
Peskin, S., 'The changes of time', unpublished MS, IGMT library, 1945.
—— 'Early recollections of Coalbrookdale and some of its villages and incidental events', unpublished MS, IGMT library, 1948.
Pevsner, N., *Pioneers of Modern Design: From William Morris to Walter Gropius*, Harmondsworth, Penguin, 1975.
Pickering, R., 'The Dunge Brick and Tile Works', dissertation, Ironbridge Institute, University of Birmingham, 1988.
Pierson, W.H., 'Notes on early industrial architecture in England', *Journal of the Society of Architectural Historians*, 1949, vol. 8, pp. 1–31.
Plymley, J., *A General View of the Agriculture of Shropshire*, London, Board of Agriculture, 1803.
Pooley, C., 'Choice and constraint in the nineteenth-century city: a basis for residential segretation', in J.H. Johnson, and C. Pooley, eds, *The Structure of Nineteenth-Century Cities*, London, Croom Helm, 1962.
Prestwich, J., 'On the geology of Coalbrookdale', *Transactions of the Geological Society of London*, 1836, series 25, pp. 413–95.
Pyne, W.H., *Microcosm: Or a Picturesque Delineation of the Arts, Agricultures and Industries of Great Britain*, reprinted 1974, Luton, Luton Museum, 1808.
Rackham, O., *The History of the Countryside*, London, Dent, 1987.
Raistrick, A., *Dynasty of Ironfounders*, London, Longmans, Green & Co, 1953.
Randall, J., *The Clay Industries Including the Fictile and Ceramic Arts on the Banks of the Severn . . .*, Madeley, The Salopian and West Midland Journal Office, 1877.

Randall, J., *Broseley and Its Surroundings*, Madeley, The Salopian and West Midland Journal Office, 1879.

—— *A History of Madeley*, reprinted 1975, Shrewsbury, Salop County Library, 1880.

Richards, J.M., *The Functional Tradition in Early Industrial Buildings*, London, Architectural Press, 1958.

Rimmell, L., 'An archaeological and historical study of the eighteenth and nineteenth century pottery industry at Jackfield Shropshire', dissertation, Ironbridge Institute, University of Birmingham, 1983.

Roberts, B.K., *Rural Settlement*, Basingstoke, Macmillan Education, 1987.

Roger, R., 'Rents and ground rents: housing and the land market in nineteenth-century Britain', *Journal of Historical Geography*, 1979, vol. 4, pp. 129–44.

Rowlands, M., *Masters and Men in the West Midlands Iron Trades before the Industrial Revolution*, Manchester, Manchester University Press, 1975.

—— 'Society and industry in the West Midlands at the end of the seventeenth century', *Midland History*, 1977, vol. 4, pp. 48– 60.

Rowley, T., *The Shropshire Landscape*, London, Hodder & Stoughton, 1972.

—— *Villages in the Landscape*, London, Dent, 1978.

Royal Commission on Historical Monuments, *Workers' Housing in West Yorkshire 1750–1920*, London, HMSO, 1985.

—— *Rural Houses of the Lancashire Pennines 1560–1760*, London, HMSO, 1986.

Rule, J. *The Experience of Labour in Eighteenth Century England*, London, Croom Helm, 1979.

—- *The Labouring Classes in Early Industrial England*, London, Longman, 1986.

Savage, R.F. and Smith, L.D., 'The waggonways and plateways of east Shropshire', thesis, School of Architecture, University of Birmingham, 1965.

Scola, R., 'Retailing in the nineteenth-century town', in J.H. Johnson and C. Pooley, eds, *The Structure of Nineteenth-Century Cities*, London, Croom Helm, 1982.

Shaw, G. and Wild, M.T., 'Retail patterns in the Victorian city', *Transactions of the Institute of British Geographers*, 1979, vol. 4, pp. 278–91.

Shaw, S., *History of the Staffordshire Potteries and the Rise and Progress of the Manufacture of Pottery and Porcelain*, Hanley, Simeon Shaw, 1829.

Shropshire Archaeological and Parish Register Society, *The Shropshire Hearth Tax Return of 1672*, 1949.

Simpson, M.A. and Lloyd, T.H., eds, *Middle-Class Housing in Britain*, Newton Abbot, David & Charles, 1977.

Skempton, A.W. and Johnston, H.R., 'The first iron frames', *Architectural Review*, 1962, vol. 131, pp. 175–86.

Slater, T.R., 'Family, society and the ornamental villa on the fringes of English country towns', *Journal of Historical Geography*, 1978, vol. 4, pp. 129–44.

Smith, D.M., 'Industrial architecture in the Potteries', *North Staffordshire Journal of Field Studies*, 1965, vol. 5, pp. 81–93.

Smith, N., 'The Upper Forge, Coalbrookdale', essay, Ironbridge Institute, University of Birmingham, 1988.

—— 'An investigation into the commercial life, trades and craft-industries of nineteenth-century Much Wenlock', dissertation, Ironbridge Institute, University of Birmingham, 1989.

Smith, R.S., 'England's first rails: a reconsideration', *Renaissance and Modern Studies*, 1960, pp. 119–34.

Smith, S., 'New light on the Bedlam Furnaces, Ironbridge, Telford', *Journal of the Historical Metallurgy Society*, 1979, vol. 13, pp. 21–30.

—- *A View from the Iron Bridge*, Ironbridge, IGMT, 1979.

Smith, W.A., 'A Swedish view of the West Midlands in 1802–3', *Journal of West Midlands Studies*, 1970, vol. 3, pp. 45–54.

Sutcliffe, A., 'In pursuit of the urban variable', in D. Fraser and A. Sutcliffe, eds, *The Pursuit of Urban History*, London, Edward Arnold, 1983.
Swannington Festival Group, *Swannington Village Trail*, privately printed, 1982.
Tann, J., *The Development of the Factory*, London, Cornmarket, 1970.
Tarn, J.N., *Five Percent Philanthropy*, Cambridge, Cambridge University Press, 1973.
Terry, R., 'An archaeological survey of the Willey Valley', essay, Ironbridge Institute, University of Birmingham, 1989.
Thirsk, J., 'Industries in the countryside', in F.J. Fisher, ed., *Essays in the Economic and Social History of Tudor and Stuart England in Honour of R.H. Tawney*, Cambridge, Cambridge University Press, 1961.
Thomas, E., 'Origin of the Ironbridge Gorge Museum Trust', unpublished MS, IGMT library, 1982.
Thomas, J., *The Rise of the Staffordshire Potteries*, London, Adams & Dent, 1971.
Thompson, H. and Winkworth, A., *Little Dawley 1983–5*, Ironbridge, IGMAU, 1988.
Thorn, F. and Thorn, C., *Domesday Book: Shropshire*, Chichester, Phillimore, 1986.
Toghill, P., *Geology in Shropshire*, Shrewsbury, Swan Hill Press, 1990.
Toghill, P. and Chell, K., 'Shropshire geology – stratigraphic and tectonic history', *Field Studies* 1984, vol. 6, pp. 59–101.
Tolley, R.S., 'Comments on the changing industrial geography of eastern Shropshire', *West Midlands Studies*, 1972, vol. 5, pp. 1–10.
Trinder, B.S., *The Most Extraordinary District in the World: Ironbridge and Coalbrookdale*, Chichester, Phillimore, 1977; 2nd edn, 1988.
—— *Coalport New Town*, Ironbridge, IGMT, 1978.
—— *Coalbrookdale 1801: A Contemporary Description*, Ironbridge, IGMT, 1979.
—— 'Coalport Bridge: a study in historical interpretation', *Industrial Archaeology Review*, 1979, vol. 3, pp. 153–7.
—— *The Industrial Revolution in Shropshire*, 2nd edn, Chichester, Phillimore, 1981.
—— 'The open village in industrial Britain', in Marie Nisser, ed., *Industrial Heritage: Transactions of the Third International Conference on the Conservation of Industrial Monuments*, Stockholm, 1981.
—— *The Making of the Industrial Landscape*, Gloucester, Alan Sutton, 1987.
—— 'The development of the integrated ironworks in the eighteenth century', *The Institute of Metals Handbook and List of Members*, 1988, pp. 217–26, London, Institute of Metals.
Trinder, B.S. and Cox, J., *Yeomen and Colliers in Telford*, London, Phillimore, 1980.
—— *Miners and Mariners: Probate Inventories for Broseley and Ironbridge*, forthcoming.
Trueman, M., ed., *Archaeology in Ironbridge 1985–6: Work Carried Out by the Ironbridge Gorge Museum Archaeology Unit*, Ironbridge, IGMAU, 1986.
—— *Archaeology in Ironbridge 1981–5: Work Carried Out by the Ironbridge Gorge Museum Archaeology Unit*, Ironbridge, IGMAU, 1988.
Tylecote, R., *A History of Metallurgy*, London, The Metals Society, 1976.
Upton, D., 'Vernacular domestic architecture in eighteenth-century Virginia', in D. Upton and J.M. Vlach, eds, *Common Places: Readings in American Vernacular Architecture*, Athens, Ga, University of Georgia Press, 1986.
Upton, D. and Vlach, J.M., *Common Places: Readings in American Vernacular Architecture*, Athens, Ga, University of Georgia Press, 1986.
Ure, A., *The Philosophy of Manufactures*, London, Bohn, 1835, 3rd edn, 1861.
VCH, *The Victoria History of the County of Stafford Volume II*, 1967.
—— *Victoria History of the Counties of England, Shropshire Volume I*, reprinted 1968, London, Dawsons for Institute of Historical Research.
—— *The Victoria History of Shropshire Volume II*, London, Oxford University Press, 1973.
—— *The Victoria History of Shropshire: Volume X*, London, Oxford University Press, 1985.
—— *The Victoria History of Shropshire: Volume XI, Telford*, London, Oxford University Press, 1985.

Wanklyn, M.D.G., 'John Weld of Willey 1585–1665: an enterprising landowner of the early 17th century', *Journal of the Wilkinson Society*, 1969, vol. 3, pp. 88–99.

—— 'John Weld of Willey: estate management 1631–1660', *Journal of the Wilkinson Society*, 1970–1, vol. 4, pp. 63–71.

—— 'Industrial development in the Ironbridge Gorge before Abraham Darby', *West Midlands Studies*, 1982, vol. 15, pp. 3–7.

Wardell Armstrong, *Limestone Mines in the Wrekin Area: Phase 1 Report*, for Shropshire County Council, 1984.

Warnes, A.M., 'Residential patterns in an emerging industrial town', in B.D. Clark , and M.B. Gleave, eds, *Social Patterns in Cities*, Institute of British Geographers Special Publication 5, London, Institute of British Geographers, 1973.

Whitehand, J.W.R., 'Background to the urban morphogenetic tradition', and 'Conzenian ideas: extension and development', in *The Urban Landscape*, Institute of British Geographers, Special Publication no. 13, London, Academic Press, 1981.

Whitehead, T.H., Robertson, T., Pocock, R.W. and Dixon, E.L., *The Country between Wolverhampton and Oakengates Sheet 153*, Memoirs of the Geological Survey, 1928.

Wiggins, W.E., *Ancient Woodland in the Telford Area*, Telford, Telford Development Corporation, Stirchley Grange, 1986.

Williams, W.H., 'Dawley New Town historical survey: industries', unpublished MS, IGMT library, 1964.

Wood, J.M., 'Company housing for the Shropshire iron foundries', University of Liverpool thesis for Royal Institute of British Architects, 1974.

Wrathmell, S., 'The vernacular threshold of northern peasant houses', *Vernacular Architecture*, 1984, vol. 15, pp. 29–33.

Young, A., *Tours in England and Wales*, London, London School of Economics, 1776, reprinted 1932.

Bibliography of Maps

*c.*1621
The Plott of Broseley
Samuel Parsons, with amendments by Francis Langley.
SRO 1224/1/32.

1637
Plot of certaine comons lying within ye towneshipps of Posnall Bentall Bradley & Wyke in ye countie of Salop . . .
Samuel Parsons
SRO 1224/1/18

1637–8
Plott of the Commons in Broseley, Bentall, Bradley and Wyke
Samuel Parsons
SRO 1224/1/19

1658
Plot of the Bounds of the Common in Broseley
SRO 1224/1/21

1658
Map of Langley's Tenement
SRO 1224/1/33

1676
Map showing lands through which mining insets passed
SRO 3703/10

1728/65
Broseley Hall Estate Map
In two folios, with additions in 1760s and early nineteenth century
SBL

1752
Carte topographique de la comté de Salop ou Shropshire
John Roque
SBL

1753
A Plan of Coalbrookdale
Thomas Slaughter
IGMT

1780
A Survey of Caughley Estate in the Parish of Barrow and the County of Salop
Thomas Bryan
SRO 1224

1786
Plan of that part of Coalbrookdale bounded by the lands of John Powis Stanley
G. Young, Worcester
SRO Cooper & Co. 1681, Box 179

1788
Plan of a Navigable Canal proposed to be made from the Marquis of Stafford's Canal in Lillishall Parish to communicate with the River Severn at Styche's Weir near Coalbrookdale . . .
George Young, Worcester
1″: 1 furlong
Boulton & Watt Collection, Birmingham.

1805
Map referring to the annexed lease, Coalbrookdale
SRO Cooper & Co. 1681, Box 138

1808
Map of Shropshire
Robert Baugh
Reprinted 1983, Alan Sutton for Shropshire Archaeological Society, with an introduction by Barrie Trinder.

1814–15
Ordnance Survey Draft Map Sheet 213
2″: mile
SBL

1827
Map of the County of Salop from an Actual Survey made in the years 1826 and 1827
J. & C. Greenwood.
1″: mile
SRO/SBL

1833
First Edition Ordnance Survey Map
Sheet 41, Shrewsbury
1″: mile
Reprinted David & Charles, 1980.

1835
Map of the Parish of Benthall in the Franchise of Wenlock and County of Salope: the estate of Thomas Harries Esq.
Surveyed by W. Hitchcock 1800
Alterations surveyed and map drawn by A. Hitchcock 1835
3 chains: inch
SRO 3956

1838
A map of Coalbrook-Dale Ironworks in the Parish of Madeley in the Country of Salop referred to the Annexed Lease
SRO Cooper & Co. 1681, Box 138

1840
Broseley Tithe Map
IGMT

1840
Sutton Maddock Tithe Map

1844
Plan of the Parochial Chapelry of Benthall in the County of Salop
Benthall Tithe Map
6 chains: inch
SRO

1847
Plan of the Parish of Madeley in the County of Salop In three parts with Enlargements
Madeley Tithe Map
Part 1 – 3 chains: inch
Part 2 – Enlargement showing lower part of Coalbrookdale $1\frac{1}{2}$ chains: inch
Part 3 – Enlargement showing upper part of Coalbrookdale $1\frac{1}{2}$ chains: inch
SRO

1856
Shropshire Union Railway and Canal
1: 3168
SRO DR 367.

1857
Plan to Connect the Severn Valley Railway with the Shropshire Union Terminus at Coalport
1: 3168.
SRO DR 369a.

1883
First Edition
Ordnance Survey
25″: mile
Sheets 43.11, 43.12, 43.13, 43.14, 43.15, 51.1, 51.2, 51.3, 51.4, 51.5, 51.6, 51.7.

1902
Second Edition
Ordnance Survey
25″: mile
Sheets 43.11, 43.12, 43.13, 43.14, 43.15, 51.1, 51.2, 51.3, 51.4, 51.5, 51.6, 51.7.

1927
Revised Edition
Ordnance Survey
25″: mile
Sheets 43.11, 43.12, 43.13, 43.14, 43.15, 51.1, 51.2, 51.3, 51.4, 51.5, 51.6, 51.7.

1956–8
Revised Edition
Ordnance Survey
1: 2500
Sheets: SJ 6702, 6703, 6802, 6803, 6902, 6903

1961–5
Dawley Development Corporation Photogrammetric Survey of Development Area
1: 500
Sheets: 0412, 1512, 0514, 0513, 0611, 0612, 0613, 1614, 0710, 0711, 1308, 1309, 1407, 1408, 1409, 1410, 1507, 1508, 1509, 1511, 1512
IGMT

1970
Revised Edition
Ordnance Survey
1: 2500
Sheets: 6802, 6902

1978
Geological Survey
1: 10560
Sheet: SJ60SE

INDEX